Proposing a "New" Cosmology Beyond Death Science

Second Edition

Theresa Talea

Rediscovery Press

Proposing a "New" Cosmology Beyond Death Science

This book may be ordered through booksellers or by contacting via the internet:

Rediscovery Press
orders@rediscoverypress.com
Rancho Cordova, California
www.rediscoverypress.com

Second Edition.

Cover design and book images by Theresa Talea, James Macaron, and Alexander Puckett.

Cover photograph by Alexander Puckett.

ISBN: 978-0-9912540-7-1 (sc)
ISBN: 978-0-9912540-8-8 (e)

Library of Congress Control Number: 2018933615

Printed and bound in United States of America by Lightning Source®.

Contents

FIGURES

Proposing a "New" Cosmology Beyond Death Science

Introduction

This book presents a process of cosmic and early creation based upon empirical and conceptual paradigms of physics and spirit-science. Spirit-science combines consciousness and spirituality with phenomena that can somehow be measured, even if only intuitively from our place of reference.

As this is largely a conceptual paper, it briefly explores the Big Bang theory, special and general relativity, M-theory, the Higgs boson and field, multiple dimensions and universes, and more in order to gain a basic but workable model of creation organizing the levels of interplay and similarity between waves, particles, and fields. This model extends to undiscovered or "new science" spaces of energy we can sense and somewhat understand when we explore beyond currently established boundaries.

Ultimately, or rather foundationally, this model arrives at the source before creation, which is not a zero-point or field because it does not contain any universal component. If we apply a loose correlation to our respective energies, this source is most similar to a nonreactive, hydrogen-like, pre-gaseous essence. This distinct, pure essence is conscious, and it can extend its awareness to all living beings through the core aspect of our compositional templates and deoxyribonucleic acid (DNA).

This book presents a variety of viewpoints building toward a clearer picture that scientists, spiritualists, and essentially every individual can consider and further study. It additionally addresses beliefs in evolution and reincarnation, and patterns of creation including fractals to better explore consciousness and diverse energies. Spirit-science sources, commonly but not exclusively originating with otherworldly entities, are included to discern probable realism from improbable fantasy. I explain how cosmology and physics are leaning toward the direction of "oneness" claims by mainstream spirit-science; therefore, it would benefit the progression of science to examine such belief systems in order to call out potential bias.

The primary theme posits that living consciousness and eternal energy substances precede the creation of "death science" based on decay. Our immediate galaxy and universe are a struggling system that could not exist without an eternal foundation in a vast cosmology.

Big Bang Theory Conundrum

Scientific discourses question whether our universe was created through intelligent design, by a random event, or a measure of both. They typically agree that it is unscientific and unrealistic when viewing intelligence as a cognitive function of a God having the ability to design every distinctive aspect and substance from which matter can bond and function in diverse ways. However, some cosmological theories, including the Big Bang, retain an arguably similar, oversimplified unrealism in which a singular energetic entity contained and extended the properties for all creation in our known universe.

The Darwinian evolutionary theory is appealing because scientists know that whatever is created has the potential to go against the grain and take creation in another direction. Accordingly, it accepts the idea that certain bacterial and viral life forms enter our planet via frozen comets.[1] Conscious life could have originated on Earth or elsewhere, and life forms would interact and shape their growth or demise. The inclusion of chaos theory explores this apparent randomness of events while also maintaining a measure of order. Order, patterns, and even particular randomness may result from some kind of intelligence.

Can scientists adequately solve the riddle of creation by analyzing creational patterns, or should they prioritize and focus their work according to a theoretical model beginning about 13.77 billion years ago as the estimated age of the universe? Data from the Wilkinson Microwave Anisotropy Probe (WMAP), based on variations in the temperature and directional change of cosmic microwave background radiation, predicted this age with near certainty (+/- 0.4%) because of a fundamental but hypothetical premise: the universe is almost entirely flat.[2]

Currently, scientists generally support the Big Bang theory that claims the universe started with a small singularity containing every attribute of universal creation compressed into a point. This singularity arose from what may be considered as nothing, because our laws of physics break down and do not apply at its state. Then, the singularity erupted (not exploded) from within itself.

When space as we know it is produced by a sudden event, it is highly disordered and non-uniform. However, physicists such as Dr. Paul J. Steinhardt note that the universe is largely uniform and smooth. To solve this discrepancy, a modification to the original Big Bang theory states that a unique inflationary energy briefly accelerated and expanded the creation process after the Big Bang. Inflationary energy is unstable, though, so it must decay into matter and radiation. This decay occurs randomly, so some components last longer than others, thus not producing an entirely smooth universe.[3]

The Big Bang theory assumes the singularity expanded in one direction, after which inflation spread out the eruption. The theory's model of creation takes the shape of an elongated bell as expansion continues. Its model begs at least two questions that I attempt to solve in this book: Why does our universal model limit the start of creation to roughly one direction when it could potentially extend in all directions? Is our universe's creational process finite compared to a more ancient and expansive cosmos, thereby validating some contextual aspects of our Big Bang model?

According to the Big Bang theory, immediately after the singularity erupted, the surrounding temperature was about

10 billion degrees Fahrenheit, and the environment already contained neutrons, electrons, protons, and elementary particles. As the universe cooled, these particles decayed or combined. This early soup did not carry light, likely due to an overwhelming presence of dark energy and dark matter.[4]

Although the precise properties of dark energy and dark matter are currently undiscovered, physicists have some understanding about what they are, or specifically, what they are not. Dark energy is inferred as the force or substance causing universal expansion, wherein its anti-gravity and possible electromagnetic properties repel space. Dark matter is a denser substance outside the property of light that can affect observed, "ordinary" matter. Dark matter is approximately measured by comparing the effects of gravity on the light of millions of galaxies via large telescopic surveys.[5] It is projected that dark energy permeates about 68 percent of space, and dark matter constitutes about 27 percent of the universe and 80 percent of its matter.[6] These figures leave us with only 5 percent of ordinary matter including stars and planets that were later formed. Matter needs to be combined in such a way to allow light to shine through it.

The early light of the Big Bang afterglow is known as the cosmic microwave or radiation background because of its abnormally high temperature. Temperature is a measurement of kinetic energy that is proportional to the pressure it exerts. Presumably, the singularity would not lose volume by compressing to create more pressure. For the Big Bang to produce such high temperature, it already had to contain abnormal force. Typically, the largest surge of energy occurs during the supernova process when a giant star's core begins to condense under gravity, creating hotter temperatures and a temporarily expanded star before the star explodes. Although these properties were contained within the Big Bang, physicists describe the universal event as sending ripples, not a far-reaching explosion as though it was a unique supermassive supernova.

The conundrum involving all Big Bang theorists is solving how the first singularity can contain all properties of creation.

A singularity is associated with a black hole as a static point that does not expand, and there is creation outside of this singularity. The Big Bang singularity, on the other hand, has a less defined boundary and movement by which the properties can extend into external creation.[7]

Our universe is still in the process of cooling from the Big Bang, which some cosmologists say could turn into a Big Freeze. This hypothesis, also known as Heat Death, expects an end to our universe in which its entropy will increase and expand until it reaches a maximum value, leaving no more usable energy for any life to continue. If universal density is lower than the average density of matter required to halt its expansion, known as critical density, then a Big Freeze could result. Conversely, if universal density is higher than the critical density, the Big Crunch hypothesis states the entire universe will eventually collapse and become drawn back by gravity into the first singularity that can become an inconceivably supermassive black hole.[8] Gravity plays an important role in the process of universal expansion, and its strength depends upon the density and pressure of matter.[9] Because most forms of ordinary matter have low pressure, and WMAP measurements indicate a much lower universal density than the critical density, a Big Freeze is the more likely scenario.

The Big Crunch is the most obvious option combining the original singularity with the black hole singularity, as if the Big Bang involved a unique nova process that ends as a unique black hole with extreme density. The Big Freeze can also include an initial black hole singularity, but it would eventually disintegrate.

The center of the universe is the best location for a singularity to expand in every direction, but it is impossible to accurately measure from our view. Therefore, a microcosmic correlation can be the center of a galaxy, wherein many, including the Milky Way, have supermassive black holes.

When a giant star dies, its energetic casing explodes as a supernova while the force of its gravity compacts the remaining body into either a neutron body without charge or a black hole. If the remaining core of the supernova has a mass (measurement of matter in an object) greater than 2.5 times the mass of the Sun, then it will create a black hole. Its large amount of gravity causes the core to collapse into itself permanently, and only a velocity greater than the speed of light can escape it.[10]

"Newton's Law of Gravitation states that every bit of matter in the universe attracts every other with a gravitational force that is proportional to its mass," states Harvey Mudd College physicist Gregory A. Lyzenga.[11] A star's gravitational pull can only reach as far as the other object's mass, distance, and orbital period can respond. Earth as our starting reference point was measured against the Sun, where the radius and mass of the Earth along with the distance to the Sun were able to determine the Sun's mass. From there, other nearby planetary bodies could be measured. Although this reference point is a far cry from the galactic center's giant supernova, these relationships show significant boundaries between created objects and forces, suggesting anything much more powerful or supernatural than what is established may not override our basic reality.

Astrophysicists use an infrared light attached to the Hubble telescope to spot the activity at the center of galaxies. Because they see a lot of activity around a compacted mass, and the galaxy is already well formed, they generally say that a supermassive black hole is consuming material.

Gravity is the weakest force, but it is theoretically stronger than any other interaction when involving a supermassive black hole. A supermassive black hole contains several million times the mass of a typical black hole, which strengthens its gravitational pull. Albert Einstein's general theory of relativity describes how gravity can bend light and curve spacetime. When a supermassive black hole's concentrated mass and gravity come into contact with its surroundings, they transform the spatial pattern into an orbiting accretion disk. Astrophysicists observe the turbulent effects upon the surrounding gas and dust, deducing that supermassive black holes grow from their consumption.[12]

While astrophysicists have not discovered a time when our galactic center has not destroyed its immediate environment, other galaxies with central supermassive black holes show no remarkable activity. It becomes miraculous thinking to assume that a supermassive black hole significantly acts outside the properties of a typical black hole relative to scale; accordingly, this way of thinking gives it similar properties to the Big Crunch with an insatiable appetite for destruction. It is unknown how far the Big Bang theory's creation process can reach for material to disperse and create, but I strongly doubt its material can become drawn back to that initial point again in a universe such as ours, regardless of varied density and pressure, because all black holes have finite properties curbing their consumption. In addition, giving the original singularity the same property as this fictitious black hole would ultimately alter the singularity into something else, thereby changing the cosmological model.

We are viewing a galactic event that happened numerous light years ago, so the speed of light has yet to show us a potentially calm, completed supermassive black hole. When it is forming, its strong gravity pulls gaseous material around it and shoots out the remains in a vertical stream called a quasar, as seen by x-ray technology.[13] Whatever light and material does not escape will eventually be trapped within the event horizon of the hole.

Some scientists believe that galaxies were created by a large quasar of highly excited, superluminous energy. *Quasar* is the abbreviation for quasi-stellar radio sources emitting radio waves further than any other visible energy. This object or wave can carry material for universal creation, but it came into existence after the creation of a supermassive black hole, deriving its energy from "mass falling onto the accretion disc around the black hole."[14] The identification of quasars helped Big Bang theory proponents refute the steady-state theory of cosmology in which the average density of matter is an unchanging constant in a continuously expanding universe with no beginning or end.

If a pattern for creation involves quasars, then this would further correlate the original singularity to a unique black hole that is similar to a supermassive black hole. If a supermassive black hole is considered as the source for new creation, then this "new" creation is something recycled and expanded from an accumulation of energy-matter existing before the formation of a supermassive black hole. I conjoin *energy-matter* to represent collaborative energy and matter, combined and separate, which includes different forms than what we observe.

The Hubble telescope shows the universe is expanding more than it is contracting, and it is continuing to expand.[15] The Earth is located near the edge of the Milky Way and is a relative newcomer, so something was able to create it. Within the Big Bang theory, the observable energy-matter that keeps expanding is probably supernova or quasar remnants from the Milky Way's (or a nearby galaxy's) big bang, unless a remnant of earlier energy-matter remains. Closer, smaller explosions after the big bang also continue to create solar systems such as ours. The explosions and subsequent creations entail recycled energy-matter.

There is an inherent contradiction within the Big Bang theory that predicates the entire universe upon its original energy-matter that eventually dies, yet it somehow contained absolute creative potential. If everything in the universe is dependent upon an original point that could be a unique star or seed, this brings us to question whether the energetic components of the first point or singularity originated within itself, and if that point could house all universal properties. It is also wise to question whether this point could arise from nothing, or if it has always existed.

If an all-inclusive energy-matter existed in the very beginning, then this was the "God" that created us, although it has also killed us. This origin of creation entwines life and death as though one cannot exist without the other. It creates the catastrophic scenario through which civilizations must become advanced to escape the impending death of their suns and planets. This paradigm is the survival of the fittest, which is narrowly attributed to natural selection and evolution.

I do not see anything truly natural about survival mode. I see the energy-matter of life remaining as life with no compromise or degeneration. I propose that the mainstream scientific community should expand its classifications and studies to include eternal-based energies that we intellectually and intuitively sense exist, and we can potentially qualify and quantify in substance. This inclusion would provide a new science outside of current mathematical equations predicting impossible infinities within the death science paradigm based on decay, which includes fractals.

Fractals and the Fibonacci Sequence

When we view the origin of life as a type of star or field that recycles itself via creation, it is true that a pattern emerges, but the pattern is a fractal. Theoretically, a fractal is an infinite pattern that replicates itself at increasingly smaller scales. Every time a mathematical equation produces a self-similar pattern, the creation is fed back into the equation. Realistically, though, fractals in our universe are not infinite. As the branching extensions of trees and spiraling seashells exemplify, the extensions are dependent upon the original capacity of their self-sustaining structures.

New Age spirit-science texts herald the "divine and sacred geometry" of the fractal because our Earth is patterned with it. The tree's sturdy trunk connects root and branch systems that grow and multiply, and the leaves show the smallest fractal pattern in their veins. The leaves die first, then the branches and roots wither, and lastly, the staff falls over dead. The exponential sequence of growth for this tree and all other life forms that die is the Fibonacci sequence.

Figure 1 shows the Fibonacci sequence up to the number 144, and it continues onward. In the first two steps, it starts at zero and naturally adds one, but each following step adds only the last two digits, thereby consuming or eliminating the entities before them. Next to the Fibonacci sequence is the Krysthal sequence that grows with every successive number.[16]

Figure 1. Fibonacci versus Krysthal Sequences

FIBONACCI   KRYSTHAL

```
0  1                              0  1
(0  1) 1                          (0  1) 1
0 (1  1) 2                        (0  1  1) 2
0  1 (1  2) 3                     (0  1  1  2) 4
0  1  1 (2  3) 5                  (0  1  1  2  4) 8
0  1  1  2 (3  5) 8               (0  1  1  2  4  8) 16
0  1  1  2  3 (5  8) 13           (0  1  1  2  4  8  16) 32
0  1  1  2  3  5 (8  13) 21       (0  1  1  2  4  8  16  32) 64
0  1  1  2  3  5  8 (13  21) 34   (0  1  1  2  4  8  16  32  64) 128
0  1  1  2  3  5  8  13 (21  34) 55    (0  1  1  2  4  8  16  32  64  128) 256
0  1  1  2  3  5  8  13  21 (34  55) 89    (0  1  1  2  4  8  16  32  64  128  256) 512
0  1  1  2  3  5  8  13  21  34 (55  89) 144    (0  1  1  2  4  8  16  32  64  128  256  512) 1024
```

Sequences can provide baselines for creational expansion, such as when atoms replicate, but they also provide comparative ratios. In the Fibonacci sequence, when a number is divided by the preceding number, the ratio becomes slightly smaller until it approaches 1.618. This number is the golden ratio, otherwise known as phi, and it starts at number 55 in the sequence. The golden ratio entails no more actual growth; when approaching the next number in the sequence, 89, it essentially becomes a finite copy. Material built with Fibonacci mathematics can only recycle its energy as constricted "expansion" until it becomes depleted and totally dies. In other words, the creation will turn into dust. In contrast, the Krysthal sequence retains every number or stage in its expansion process, providing full integrity in its creation with open connection to its origin.

Our cosmology contains multiple levels or stages of development existing long before the Earth's creation. The Krysthal sequence would be placed in an early stage, so our particular density and diversity would entail a very large number in the sequence. To the contrary, the Fibonacci sequence started at a later stage as a copy for creation to occur in another direction. Perhaps it would be most accurate at our expanded stage to work with ratios instead of whole numbers in the advanced sequences, although we can see replicas of early sequencing as microcosms in our environment. Earthly formations are commonly built with the Fibonacci sequence and fractals, but Krysthal sequencing is still present in their foundational frequencies.

What I call the Krysthal sequence was initially introduced as the Krystal sequence from author and speaker Ashayana Deane (henceforth named Ashayana), who has thus far received the majority of her information from otherworldly entity groups named Melchizedek Cloister Emerald Order (MCEO) and Guardian Alliance (GA). The third edition of my book *Eternal Humans and the Finite Gods* reports their origins and teachings to reveal their foundational Law of One belief that dominates New Age religions.[17] I include some of their teachings in my analysis of cosmology because, although some concepts may help to push scientific boundaries, I see a growing trend in physical cosmology heading in the direction of New Age teachings, which are also represented in the estimated everything-in-one Big Bang singularity.

The first two diagrams in Figure 2 show geometric representations of Fibonacci and 12-point kathara grid spirals based upon how the MCEO-GA compare them. These spirals are not a direct comparison because the Fibonacci grid and kathara grid are dissimilar bases, and only the Fibonacci spiral incorporates its numerical sequence.

The Fibonacci spiral is drawn on a tiled spreadsheet with squares containing increasing numbers of the Fibonacci sequence. This produces a spiral with an off-center base that expands asymmetrically as it connects the corners of each larger square. The MCEO-GA do not provide a similar tiled representation of the Krysthal sequence but instead show a self-contained diagram of small-to-large grid structures called kathara grids that are connected by their apexes when turning 45 degree angles to the right, thereby forming what they deem as the "Krystal Spiral" (p. 2).[16] Their kathara grid spiral does not represent the Krysthal sequence because it expands by $\sqrt{2}$ which is 1.414 per kathara grid, not the full multiplier of 2.

Figure 2. Fibonacci and 12-Point Kathara Grid Spirals

FIBONACCI SPIRAL-1   12-POINT KATHARA
                              GRID SPIRAL

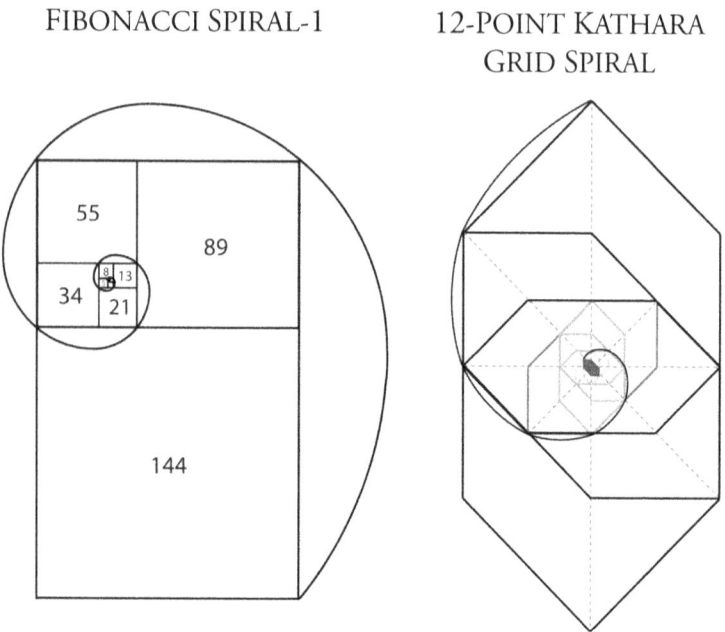

FIBONACCI SPIRAL-2

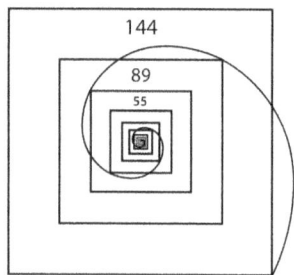

The kathara grid is the primary mathematical-geometrical backbone upon which universal and bodily matter are assembled in a patterned order.[18] The MCEO-GA teach that this grid has 12 key locations of matter that draw from a unified field of energy and share their coding throughout their respective regions, which would each entail a type of non-explosive "big bang" toward subsequent creation including stars and planetary systems. This viewpoint is a variation of the Big Bang theory that implies all matter within a large spacetime realm originates from one key location and structure. The MCEO-GA equate the kathara grid "points" to certain stars and planets in a galaxy when these numbered points actually represent unique large dimensions, which I will soon explain.

The MCEO-GA say their Krystal spiral connects each kathara grid to its center, but we can see in Figure 2 that its apex connections resemble the Fibonacci spiral in an unending search for their respective cores. The reason the Krystal spiral *appears* to be centrally based is because it is created by the rotation and enlargement of subsequent kathara grids on top of the original one, therefore basing the spiral off of a 12-point kathara grid.

The MCEO-GA teach a kathara grid constructed with 12 key points, but we can create similar spirals with 15-point and 9-point kathara grids, for example. This logic brought me to create the same arrangement with the Fibonacci spiral's squares by placing successive squares of the Fibonacci sequence over the first one in the middle, as shown in the Fibonacci Spiral-2 diagram. The Fibonacci Spiral-2 enlargement multiplier is 1.618 to reflect the golden ratio, but it would create a nearly identical pattern with the Krystal spiral multiplier of 1.414. We can probably create a uniform spiral from any replicated shape with a curved connection and a constant multiplier.

The ease of creating spirals in this manner brings me to question whether the MCEO-GA version of a Krysthal spiral is really the fully natural creation that *Krysthal* implies. *Krysthal* is the combination of the seven frequency tones Ka Ra Ya Sa Ta Ha La, each coming from an early level of creation with eternal stars called krystars,[19] most of which have planetary living conditions, which I later explain in this book along with early creation domains. *Christ* is a distorted galaxy-based variation of the root word *kryst*.

Spirals involve energy circulation that can precede structure, and they can vary in mathematical sequence and construction. Regarding a truly Krysthal spiral, it gains particulate streaming elements in its process of expansion, coming together in an ordered sequence to form a precise path. To arrive at this understanding, along with my intuition, I questioned what I elucidate in "The Origin and Expansion of Creation" section as the origin of energy consciousness, called the All That Is, The Pure Essence—an eternal, conscious essence that can connect with created realms and entities of energetic integrity and pure intent. (I often write it as ATI,TPE for brevity, although an abbreviation is not preferred by the All That Is, The Pure Essence in direct communication with it.) It states, "The Krysthal spiral is symmetrical and perfectly sequenced according to a growth pattern."

The Krysthal spiral originates near the beginning of creation, and it provides energetic frequencies for the progression of creation into eternal domains and realms. This entire energy spiral aligns with one's center or core essence, and it involves the Krysthal sequence of expansion, confirms the ATI,TPE. The Krysthal spiral is totally symmetrical because its frequency pattern is complete and unbroken. It is unlike our spiral models siphoning energy into a finite structure that appears uneven when cut in half. The fully natural spiral may resemble a string filled with energy that multiplies outward in symmetrical wavelength sequences.

A spiral based upon kathara grid structures is not the same as the full Krysthal spiral of energy circulation. The MCEO-GA perspective of creation provides building blocks toward more knowledge, and they can be valuable when compared to science as we know it; however, something is amiss in their information, so I aim to show why.

The universe that we can measure is a fractal universe. Its alleged Big Bang entails death as well as life, but life precedes death. If we stretch our minds a little bit, we can deduce that the fractal pattern in our universe mimics another earlier pattern of full integrity. This means our universal science not only falls short of greater cosmological mechanics, it also can misrepresent them due to its fragmented nature.

Natural kathara grids contain self-sustaining, ordered patterns of matter and pre-matter that involve the cores of their existence unlike fractal patterns that emerge from the last two Fibonacci creations. Some of the natural pre-matter are sound and light units called keylons. The MCEO-GA call their conjoined teachings *Keylontic Science* because it is based on these keylons that contain living codes of "geometric-electric and magnetic structures that create the foundations for all form and structure within the dimensional systems" (p. 25).[20] I will call natural, eternal science *Krysthal Science* because it defines almost every layer of eternal creation, including those built with keylons, and it is inherently compatible with foundational aspects provided by earlier realms.

The MCEO-GA place the kathara grid and other geometries within a spherical, star-like realm.[21] If anyone, scientists included, believes that the entire universe was created by a unique type of star, then black hole mechanics may not be entirely different from the MCEO-GA's Keylontic Science. If black hole mechanics are distortions or variations of Keylontic Science, then could Keylontic Science be a distortion or variation of an earlier science such as Krysthal Science?

Every point on a truly natural kathara grid is filled to its full energetic integrity, which is eternal energy-matter. If material is pulled off of those original structures to create lesser copies, then these copies can contain increased amounts of distorted "junk" and compaction toward death if they are eventually cut off from their energy source. In addition, partial circulation or energy flow that recycles a structure's energy-matter has a limited life span.

An example of partial circulation is the torus, which is "the surface generated by constructing a tube around a circle."[22] The MCEO correlates the popular Chinese Taoist Yin-Yang symbol to dual torus-shaped, electromagnetic fields revolving continuously. Both black and white contrasting portions of the symbol show a single hole of the torus to prevent the structure from shrinking to a point. The torus fields act as electromagnetic harnesses that influence the Earth's axis, causing the Earth's wobble. The harnesses create a closed-loop, dual pathway of energy erupting through the North and South Pole, appearing as a bulbous, "'Poison Apple' geomagnetic field and magnetosphere" (p. 19).[23] Conceivably, the Earth's energy would radiate in all directions with no erratic wobble once it is free to fully "breathe." Distorted displays of energy-matter, including stifled energy circulation, are the creative expansions of fractals, which are not truly creative.

Fractals are commonly derived from a living source in an original state, whose stage of existence would not contain the actual fractal. The fragmentation of an original material is what starts the snowball "creativity" of the fractal that partially grows in subsequent stages until it cannot hold any more form. In contrast, the original material regenerates itself because it has life-giving energy in every layer.

Reincarnation versus Evolution

Fractals help create reincarnation. Reincarnation does not simply provide a new physical body; it is a recycling process that reduces a measure of natural energy in the individual's departed composition, which can translate to reduced energetic integrity and capability in the new human body.

Dimensional locations near Earth contain different amounts of natural energy, and some—which can be known as review planes—provide the personal space to evaluate one's earthly experiences. However, there is a proportional relationship between the degree of energetic distortion and difficulty of self-realization. When Person A enters a more natural review plane, this person has a choice whether to reincarnate back on Earth with the intention to make a positive contribution. The increased natural energy brings partial regeneration, which may not allow the process of reincarnation to dig in as deeply as it would without any regeneration.

A lesser amount of regeneration occurs in the more distorted review planes to those who positively alter their life's course; if they choose to reincarnate, they will carry this small measure of regeneration with them. The various review planes are only some of several dimensional holding places for departed Earthlings, confirms the ATI,TPE.

Alternatively, reincarnation can forcefully occur when people submit themselves to their religion or spiritual philosophy, effectively entrusting an external agent, such as a god or geometric entity, to decide their course of action for them. When Person B has already relinquished one's individuality and choice, this may direct the person to a more fragmented dimensional location. Person B may be forced to reincarnate when subjected to the power of that unnatural region, or when one's compositional essence does not have enough natural energy-matter to strengthen the person after death.

When uninterrupted reincarnation cycles occur, the broken ties of fractal expansion cannot hold the person's form after seven successive times. This amount of time corresponds to the stages of the Fibonacci sequence immediately prior to the golden ratio when starting the process of continuous reincarnations at number 2, as shown in Figure 1. While the Fibonacci sequence is technically "natural" on Earth, its ratios can and sometimes do apply to the increasingly fragmented progression of forced reincarnations, confirms the ATI,TPE. Regarding the departed humans who do have a choice to reincarnate, most of them choose not to do so after seven times because of the fragmentation, loss of memory, and dangerous risks involved, states the ATI,TPE.

When pondering ways out of a possible reincarnation trap, we may consider the topic of evolution. The conventional scientific theory of Darwinian evolution states that a basic life form "mutates" in growth to become advanced. Its advanced DNA (deoxyribonucleic acid) then passes on to countless generations after death. Over a period of time, later replicas with adaptive experiences gain an upgraded genetic template that has learned to master its environment. Within its own life, the organism can also evolve in a small way, such as a bacterium that becomes resistant to an antibiotic. When applied to cosmological creation, this theory would presume the universal singularity was a multifaceted but unintelligent, unconscious substance that expanded as building blocks upon which consciousness and intelligence somehow emerged.

Darwinian evolutionary theory contains notable paradoxes. It appears to illogically derive consciousness from no preceding measure of consciousness, as if something comes from nothing. In addition, it denotes a finite nature in the original substance due to its creation of death mechanics, so the recycled universal fractal will inevitably stunt intelligence and therefore evolution, thus validating reincarnation. The conceivable recourse is to place conscious intelligence within the "genetic" material of what existed originally.

A non-conventional view about evolution states that the life form is eternal, and it will evolve from its starting point that is very comfortable and aware. This life form never started out as a bacterium or amoeba but was already naturally advanced. If this being is a human, for example, our entire genome would be fully developed and utilized, so evolution would entail reaching other realms or dimensions. This type of evolution will increasingly gain non-fractal energy-matter to eventually enter fully Krysthal systems that would expand our intrinsic genetic capacity toward its full potential. It involves order based upon a natural kathara grid where the lower dimensional person will fulfill that template in a stair-step process of ascension. No part of oneself, especially the body, is lost in death or fragmentation.

Another non-conventional view is most applicable to our current human condition. We contain a mixture of a fully functioning, eternal template and a partially functioning, finite template in our DNA. Evolution would progressively heal our finite genetic distortions to increasingly reveal us as highly capable, eternal Humans.

I will illustrate how reincarnation reduces evolutionary advancement in a practical story involving the religious perspective. When someone is brand new, and then evil befalls him, the purity is obscured under a layer of confusion about why this would happen to him. There is always a first life before the process of reincarnation can occur, but we rarely if ever hear of this time from the religious reference, only the later lives of lessons learned, forgotten, and re-learned. When the world is as it is with all its evils in place, of course the brand new, innocent being does not deserve the powerful acts against it. That in itself creates pain; to say this person deserved it because of past experience or karma is simply untrue.

So, the innocent person was hurt, but his religion does not divulge correct information to support his innocence. Instead, this person may be shunned or punished for genuinely reacting to the evil, so he grows up feeling as though he is a bad person. This treatment causes a change of perception and behavior toward the evil by which he reasons it was acceptable, and he even deserved it. This dumps filth upon the pure self, so the person dies a little inside. He was not valued for being pure, so he now thinks, "What's the point? This world is confusing and unfair, so maybe I should conform my sense of self to these judgments." His goodness then dims, and his actions eventually confirm his religion's assertion that he is at fault and unevolved. Then, each subsequent life would likely be filled with more pain because of more systemic punishment. Even if the person is eventually punished correctly, he will not grow much from it because of the initial unjust and unrectified blame.

My mind became opened to the reality of reincarnation when reading stories about children who clearly recount their previous lives. Interestingly, some of them were born with a specific pain or blemish, and it directly relates to a serious wound that occurred in the past life. For instance, a boy named Chase remembers being an adult soldier who was shot in his wrist. Since birth, he had severe eczema on that precise spot. Whenever he felt troubled, he would scratch his wrist until it bled. Recalling his past life, he knew he did not want to be in that war. When he faced his emotions about the war, his eczema healed within a few days and never returned.[24]

We can figuratively correlate our origin and potential

energetic evolution to the Krysthal sequence. As a brand new existence, we would equal the second number 1 in the second line of Figure 1's Krysthal sequence. Our individual wholeness includes a connection to the Source at 0 and a measure of the self-regenerating energy expansion before us as the first number 1. In the infinite nature of life, we can change to a lighter density or achieve greater self-actualization, so our next energetic transformation would make us number 2. The next one after that becomes 4, and so forth. This model expands with fullness, not taking anything away from what was there before.

In contrast, the Fibonacci sequence can directly correlate to the forced reincarnation process. In Figure 1's Fibonacci sequence, we contain the same individual wholeness as the second number 1 in the Krysthal sequence, but when the human body dies and may reincarnate to number 2, this process will somewhat hinder our connection to the Source. The next reincarnation will cut more deeply into the natural capacity of the body and diminish it as number 3, which is half the potential it could have been as Krysthal number 4. To exemplify this phenomenon, the myriad of negative and stressful experiences of the previous human life can become imprinted in one's energetic structure that returns for the next human experience and stunts its growth. The following few lives grow fractionally, but each successive life grows less than the one before it. The seventh number in the sequence after our birth is 55, which opens the door to the static golden ratio.

The fractal model can never approximate a fully aware self, yet the religious view about reincarnation requires us to be highly enlightened people so that we can become Ascended Masters beyond Earth. How do we properly evolve if we simply cannot evolve much or at all with fragmented energy-matter? This is why some religious models include countless reincarnated lives: to give the idea that we can eventually overcome our "sins" of death science, when reincarnation does not actually change the mechanics of death. Continuous reincarnations do not allow proper growth of energy in a 2:1 ratio; they put us in a prison of recycled energies that can eventually fizzle out.

Another spiritual belief in reincarnation states that we were angelic beings before this life, and we elected to incarnate here for the sake of experience or a mission. Incarnation in this situation entails complete human embodiment. A lighter density "person" does not completely enter a denser body by covering itself with that other body; the less dense being already has its own material body. Entering a functional, denser body is an act of possession and potential assimilation of that energy-matter. If an angelic being with more natural energy-matter than ours as an earthly species wants to maintain its energetic integrity, then it arguably would not compromise itself nor resonate with increased death science, therefore not intermingling with our matter.

Science and religion alike are confused about the combination of life and death in our realm. Scientists may perceive the Big Bang as a point of energy creating our entire living universe, but the inevitable death of that creative point nullifies the eternity of life. It is arguably impossible for science to deduce eternal life within the current framework of recycled energy-matter. Religion agrees with science about the circle of life and death; however, religion believes that life is either embodied or disembodied by something beyond us, and we must become transformed by its power to reach its ultimate state. Both schools of thought leave nearly everything out of our grasp. New Age spiritualism aims to remedy this disempowerment by saying we are fundamentally gods who create and control essentially everything in our lives, but its popular spirit-science simultaneously promotes the golden ratio fragmentation, therefore proving its fantasy. Our current reality has us live in the gray area. This gray area maintains creative life, and the measure of influence we have over our lives can be increased the more we realize how multifaceted and connected we and the cosmos are, separately and together.

Probing into the Composition of Other Dimensions

Scientific theories of the Big Bang and evolution have

traditionally believed that all matter essentially coexist in a vast three-dimensional space. The factor of time or cause and effect has allowed the initial singularity's composition to change since producing the Big Bang, and the universe to evolve into more space with diverse creations. The way we might regain the state of origin is if we somehow reverse time and everything that occurred with it, which is unrealistic.

Alternatively, we could incorporate at least one extra dimension to allow for the pre-Big Bang condition to exist simultaneously with our present universe. Extra dimensions in this context would operate independently from our three-dimensional experience in which one dimension is a coordinate denoting a line in space. I will introduce another type of dimension, called a large dimension, which arguably provides another reality than our experiential, universal fabric that is inherently interwoven with time.

Albert Einstein took Cartesian coordinates (x, y, z) further to include another dimension relating to time (t), denoting every four-dimensional coordinate system as a "point-event" of spacetime (p. 151).[25] Time is not a spatial dimension; it is a scalar quantity defined by a magnitude or numerical value indicating a one dimensional measurement. When time is added to three spatial dimensions, they create a continuum having characteristics of a manifold and field.[26]

Einstein's equations sparked inquiry into the great debate about what came first: a particle, a field, or a cosmic "egg" containing both and much more.

Particles are not necessarily simple objects. The principle of wave-particle duality states that particles can sometimes be waves, and waves can sometimes be particles; however, both cannot arise at the same time when provided two paths as two slits cut into a barrier. This principle has thus far proven true in laboratory experiments wherein "all relevant subsets of an ensemble must be sampled with equal probability" with unbiased measurement settings.[27] The wave-particle duality observation applies to electrons, photons, protons, and neutrons, although protons and neutrons are less detectable with shorter wavelengths due to heavier mass.

As so much of existence is beyond our observational skills, I think it is possible that a portion of the agitated particle can disseminate into a different type of wave that is undetectable to our current technology. The study into less material energies and densities could potentially reveal a dual existence as both particle and wave in slightly different locations. This would show a multifaceted ability and potential consciousness constructed into the particle that transcends some distance when its wave attributes are extended, divided, and/or displaced. These simplified waves or flows are not necessarily wavelengths or frequencies; they could be foundational aspects that developed according to a sequence, which would potentially reframe what we consider as elementary particles or waves.

In 1905, Einstein published papers introducing the special theory of relativity (special relativity) and his famous equation $E=MC^2$, which defines energy as mass multiplied by the speed of light squared. He proposed that light exists as tiny packets, called photons, which act as both particle and wave at different times because of special relativity properties accounting for their movement at the speed of light at any position in space. Accordingly, this principle applies to other quantum wave-particles traveling up to the speed of light when they maintain uniformity in their different coordinate positions.

Special relativity incorporates speed (measure of traveled distance divided by duration of journey) with the concepts of space and time, and it posits that space and time adjust themselves to keep the speed of light fixed. Einstein postulated that the speed of light is a constant inside and outside of a vacuum, but now it is viewed as a limit.[28] Photons have zero rest mass in a laboratory vacuum while traveling at the speed of light. They are always moving, so they can be measured in terms of their electromagnetic waves called quanta.

The main obstacle to special relativity becoming a law is gravity. Gravity is the most difficult force to understand; for instance, the Earth's gravity does not override the weak pull from a magnet to a paper clip.

A large object's gravitational pull is instantaneously realized by surrounding bodies, so this immediate change exceeds the speed of light. To solve this conundrum and expand the special theory of relativity, in 1916, Einstein published "The Foundation of the General Theory of Relativity" identifying spacetime as the medium transmitting gravity.[25]

The general theory of relativity states that the spacetime continuum becomes curved by its interaction with objects, especially ones that are very heavy and large, and it is this curvature that causes gravitational waves, which in turn affects nearby regions.

Particle physicists, including Einstein, have generally believed this interaction with dense objects causes fields and forces to arise. For example, the mass of two planets would involve or even cause gravity to turn the space between those objects into an elliptical orbit. Another example would say a magnet creates a magnetic field that attracts iron. To the contrary, quantum field theorists say the magnet interacts with the already present magnetic field and alters the local conditions of the field.

According to $E=MC^2$, an object's energy is reduced when its mass is reduced. I propose that a wave-particle can exude greater energy with very little mass, irrespective of how far it travels. My hypothesis extends to a cosmology beyond our current scientific method based on measurable quantities, or observables, primarily involving objects and particles. But when saying an observable particle acts as a wave, this wave can either be a quantum function of a microscopic string or a perturbation of a field.

String theory states that fundamental universal objects are not point-like elementary particles; instead, each one is a string spread out in one dimension. Vibration and oscillation of the strings give the different particles we can see. In contrast, quantum field theory unites quantum mechanics and special relativity to state that fundamental universal objects are quantum fields acting as operators able to create or destroy particles.[29]

It is debatable whether some or none of these particles are

indeed isolated particles, for they can be merely localized points in fields. In the words of unified field theory physicists Horst Eckardt and Laurence Felker, "All physical mass-points of a field theory are actually densities—i.e. quanta of matter-energy spread over a volume of space" (p. 6).[26] Fields are voluminous spaces containing multiple points as localizations, densities, and excitations that move in one or several directions, or with no discernable movement. Every point in the space of a field has a value, such as temperature.

In the study of mathematics before Einstein, "Maxwell's four equations describe the electric and magnetic fields arising from distributions of electric charges and currents, and how those fields change in time," states University of Virginia physicist Michael Fowler.[30] Maxwell's experiments revealed that electric and magnetic fields can exist on their own, potentially far away from their charged source, because their field interactions can self-sustain. This adds complexity to the discussion of whether a wave or field, or both—theoretical physicist Dr. Sean Carroll states they are essentially the same thing—came first.[31]

Some fields, including electric and magnetic fields, can gather pre-existing particles or waves (however they are defined) and start a new generation of creation. These occurrences can support the hypothesis of a kathara grid galaxy upon which denser objects are formed, and a different type of universal space may exist prior to the kathara grid. When reducing fields to geometries, physicists mainly utilize Euclidean geometry with planes and solids, typically by measuring mass-points with straight lines. The kathara grid fits this geometry in which there is implied multidirectional volume. As this topic becomes purely conceptual, I must take several steps from what our current science has proven to arrive at a new model about the composition and process of creation that reaches to the very beginning of existence.

Scientific reasoning can be inductive and potentially unrealistic when desiring to turn cosmological hypotheses into a unified theory or law. Cosmology involves many variables and experiential realities, so it is reasonable to presume in our chaotic universe that several laws coexist with some discordance instead of one supreme law generating everything.

One so-called theory of everything, known as M-theory, is a notable idea in high energy physics originating with the exploration of strings, otherwise known as filaments. M-theory unites five different, healthy theories of strings that consistently account for quantum gravity in an 11-dimensional format. String theories arose from the advancement of Einstein's work to include additional dimensions containing the four physical forces— strong nuclear, weak nuclear, electromagnetism, and gravity—and all particle matter. When more dimensions were mathematically opened up, strings became increasingly stable and were therefore labeled superstrings, an abbreviation of supersymmetric strings.

Supersymmetry provides a partnership between fermions and bosons—families of the observed elementary particles or fields—that allows them to transform into each other and mix. Fermions have one-half integer spin, and they compose matter and take up space. Identical fermions cannot occupy the same space and energy state because their electrons repel each other. In contrast, bosons have one integer spin, and they create or carry force. Identical bosons condense together in the same space and energy state.

Supersymmetry assumes "utmost simplicity and symmetry" in the beginning stage of our universe whose perfect symmetry became broken and complex as the universe expanded and cooled.[32] While I agree with the assertion of early cosmological simplicity with some facets of symmetry, the principle of supersymmetry relies upon hypothetical shadow particles called superpartners with significantly larger mass that are hidden, possibly in compacted dimensions, and give rise to our observed particles. High mass bosons would be superpartners to less massive fermions, and high mass fermions would be superpartners to less massive bosons.

Superstring theories involve our four dimensions of spacetime and six extra spatial dimensions, commonly stated as compactified. String theorists often use a circle, or torus, to represent a compactified dimension. The six compactified dimensions may be curled up together at every point in space, forming twisted Calabi-Yau manifolds. Manifolds are topological spaces in which the neighborhood of each point has the ability to morph into a Euclidean space.[33] When compared to quantum field theories, Calabi-Yau manifolds are similar to localized points of a field because they are mapped on an extremely small-scaled grid and essentially constitute that grid; however, their multiple dimensions do not define typical fields. At best, these supposed higher dimensions would be either a background or interwoven quasi-field to ours whose components must translate into our observable components.

When strings interact exceedingly strongly, an additional dimension opens up, which M-theory includes as the 11th dimension.[34] M-theory incorporates types of membranes called branes, which can be multi-dimensional mediums exhibiting their respective number of dimensions as objects of varying sizes. For example, a three-dimensional brane could be a large box, and a 10-dimensional brane could be a hyperplane-like object, which is a subspace of one dimension less than its ambient space. The hyperplane-like object is a space-filling brane equal to the total number of spatial dimensions throughout an entire spacetime, which is another option for the three-dimensional brane.[35]

As the "M" in M-theory is unknown, it may refer to a fundamental two-dimensional membrane as an extended object whose one-dimensional portion wraps around the compactified 10th dimension, and the other portion extends into the 11th dimension.[36] M-theory does not have strings in the 11th dimension; the 11th dimensional brane gives rise to strings in the lower dimensions. This object and dimension, as well as the other extra dimensions, may exist at a larger scale than quantum strings.[34]

String theories are perceived to unify the four physical forces because they not only include gravity but also attempt to define it. M-theory asserts that gravity appears to be the weakened force because the graviton wave-particle freely travels between branes, while our matter and strong forces stay on or in our three-dimensional brane. Matter that is strongest in our experience is comprised of open ended strings that tie down to our brane. The graviton, on the other hand, is a closed loop string that is free to travel.

Some M-theorists posit that we can feel other brane dimensions through escaped waves such as gravity. More boldly, they propose it is the interaction of branes—most likely a unique 11th dimensional brane or 10-dimensional hyperplane-like object with a very similar one of another universe—that created the Big Bang, which would not be so big after all in a greater cosmos. This ultimate brane would not be an object barely larger than strings; it would be a large brane world, or more appropriately, a large brane universe. This idea assumes a volatile cosmos with parallel, floating membrane planes that collide and ripple out destruction and creation. The ultimate brane stretches in a non-uniform capacity until its creative, energetic potential is reached and cooled, and then it pulls back into a plane in a Big Crunch.[37] I envision a cosmos much more stable than that, where it is not bound to a type of cosmic plane or egg that has power over all of its creation.

Some physicists are exploring the concept of large extra dimensions to explain the weakness of gravity, which could align with an M-theory based proposal of an extraordinarily large brane world or universe as well as other large branes similar to ours. "In brane-world scenarios, matter fields and force particles (except for gravity) are confined to exist on a 3+1 dimensional 'brane' (membrane) embedded in a space with large extra dimensions," explains California Institute of Technology research physicist Leo C. Stein.[38] Brane worlds or universes can include different scales for dimensions if the 11th dimensional brane is the backbone to our large brane, and the small and possibly compactified other dimensions exist in-between or within these two branes.

There may be two or more entirely different scales of creation at play in our universe. For instance, the quantum scale can apply to a building block of our particular matter, while a very large scale can apply to a narrowing or step-down process—that does not necessarily imply division—from higher, large "extra" dimensions to ours. The tiny-to-larger scale could appear as a funnel or acute angle shape that is conjoined with an inverted funnel or inverted acute angle of the larger-to-smaller scale. These scales also allow room for other potential scales before them that can provide a larger and somewhat different spatial image.

When extended to a large region in the universe, a distinct space can exist independently from other spaces, and it may also be a subspace with inherited operators while containing individual quantities. Simplified reasoning can assume that all forces and properties of matter exist in the same space, as the New Age Law of One belief fundamentally states; however, vectors (quantities with direction and magnitude) and dimensional measures recognize different locations in space from one point to the next. The reality of points existing in separate measurable spaces means that *space* can be a subjective term, so this gives caution to how we identify dimensions.

I define a large dimension as a unique space with a unique frequency or frequency blend where material creation is represented somewhat differently. The space of one large dimension contains a three-dimensional reality of geometry such as ours that includes time. I define time as the momentum of particle pulsation in a large dimension of a harmonic universe, which I properly introduce soon.

M-theory is based upon strings of the Planck length that vibrate at fundamental resonant frequencies. Different resonant frequencies can determine different fundamental forces.[39] The Planck length is equal to 1.616252×10^{-35} meters. This is almost identical to the golden ratio value of 1.618 but on a much smaller scale. M-theory is another theory of recycled and finite energy! According to the Swinburne University of Technology:

> The Planck length, and associated Planck time, defines the scale at which the currently accepted theory of gravity fails. On this scale, the entire geometry of

spacetime as predicted by general relativity breaks down. The main reason for this breakdown is that the Planck scale is smaller than the quantum wavelength of the Universe as a whole.[40]

In Einstein's day and throughout the 20th century, the Planck scale has defined the smallest measurable quantity in our known universe. The Planck scale is a high energy scale at which gravity becomes strong and possibly comparable to the other forces. I wonder if we can extrapolate this mechanism to a large scale to help us understand black hole mechanics in terms of extreme density and gravity, and the golden ratio, so we can facilitate a beneficial approach in dismantling the death science mechanism and ideology.

Physicists are discovering tiny particles with particle acceleration technology such as the Large Hadron Collider (LHC), which smashes atoms at approximately the speed of light in a large facility called the European Organization for Nuclear Research (CERN) in Geneva, Switzerland. The main objective for utilizing the LHC is to find the Higgs boson, nicknamed the "God particle" because it could prove how particles gain mass. The first three-year run, which ended in February 2013, gathered enough data for CERN scientists to determine that they found a particle fitting the description of the Higgs boson; however, research may be inconclusive as to whether it was the fundamental boson under the Standard Model of particle physics.[41] It is possible that this Higgs boson-like particle "could be a composite particle made up of two even smaller techni-quarks, bound by a theoretical 'Technicolor' force" that is stronger than our currently known forces and can break electroweak gauge symmetry,[42] which causes measurable quantities of a field configuration to remain unchanged when the field locally transforms.

Contrary to other bosons, the Higgs boson does not create a noticeable force. It is probably derived from an excitation in the universal Higgs field that imparts mass to all other particles. Although the Higgs boson essentially completes the Standard

Model of particle physics, the Standard Model is incongruous with a field origin and gravity, which the potential discovery of the graviton, a possibly massless boson, could further clarify.[43]

Physicists are open to discovering a new physics if the Higgs boson is found to have less mass and higher frequency than the Planck scale allows. This new physics could merely involve particles in the electroweak scale that have less mass than the Planck scale and can exist outside of gravity, or it could be a really new physics that can effectively explain dark matter and dark energy as well as the luminous, massless boson, the photon. Accordingly, opening up scientific possibility toward new mathematical scales, such as an expanded Krysthal scale allowing other large dimensional "branes" similar to ours instead of tightly coiled spaces based on fractal decay, would involve not only innovation but also intuition toward determining the differences between death and eternal sciences.

I envision several different, new physics allowing variations of energy-matter both seen and unseen along the life-giving spectrum of various systems, including the existence of earlier generations of particles and fields, or pre-particles and pre-fields. Further exploration of the Higgs field and mass-giving mechanism can potentially reveal such pre-particles and pre-fields "on the other side." Massless particles such as the photon are not affected by the Higgs field, which gives the possibility that these particles may exist prior to the formation of the Higgs field. For particles to gain mass, a phenomenon such as quantum tunneling can draw from particles' environments and give them a burst of energy to push them through the Higgs field "wall."[44] This means that the Higgs field can act as a barrier, and a conjoined, effective force or force field (outside but adjacent to the Higgs field) could cause particles to gain mass when passing through the Higgs field. Quantum tunneling can also suggest a step-down and step-through process of large dimensional creation.

In 1971, theoretical physicist Claud Lovelace published the first string hypothesis with several extra dimensions—26 in total—to overcome the emergence of tachyons in bosonic string interactions with closed-string loops. Lovelace knew his progressive idea would not be taken seriously, but he submitted the paper anyway. Although his hypothesis erroneously predicted the existence of only bosons, its introduction of several extra dimensions helped pave a way for theoretical physicists to develop superstring theories.[45] Tachyons, however, are yet to be taken seriously as subatomic particles moving faster than light.

In quantum field theory, particles must be massless to travel at the speed of light. This would place the tachyon in the "imaginary mass" realm where it would be negatively numbered, non-local, and impossible for us to measure.

Strings called light strings are massless at the speed of light. "In a sense it becomes a generalization of a ray of light, a ray that can vibrate and spin," states physicist F. David Peat.[46] The massless string can act similarly to a particle with mass, but it would operate independently from the types of fields we have studied. The light string could create and accumulate particulates of matter from a magnetic force outside of an electromagnetic field. These particles then line up in an orderly fashion to be carried upon the string's wavelength.

Peat writes in *Superstrings and the Search for the Theory of Everything*:

> In singing, the higher you go up the scale, the more energy you need to produce the notes. In an analogous way, the quantum notes of the string—its quantized vibrations and rotations—are steps in a ladder of energy.[46]

Matter becomes less dense and filled with more energy as it increasingly enters lighter and higher realms. The music scale is an excellent analogy to extrapolate vibratory and spinning strings to multiple dimensions. It can also illustrate the figurative spiral staircase made of older energy components that allow vibratory objects to exist within resonant large dimensions, which provide energetic steps.

Have you noticed a constant humming in your ear in the dead of night? This is not to be confused with tinnitus; it is the environmental and internal energies with which we interact. In the following excerpt, composer Susan Alexanjer explores the vibrations made by our own bodies. Our DNA has a string-like template as a double helix that receives and transmits electricity.

I proposed that we try to measure the actual molecular vibrations of the bases that make up all of DNA as we know it, as it appears in all life forms. To my astonishment, Dr. Deamer [cell biologist] explained that the vibrations were easily measurable, using an infrared spectrophotometer. By exposing each base to infrared light and measuring which wavelengths each base absorbs, it is possible to identify a unique array of approximately 15 different wavelengths for each base. Since each base has a slightly different atomic structure, it will vibrate in a unique manner. As the atoms of carbon, hydrogen, nitrogen and oxygen receive the light, they absorb some of it, depending on their vibrational frequencies, and those absorbances can be measured, plotted on a graph, and read as numbers. These numbers, in turn, represent a wave-length "scale" on the light spectrum, but very fast, very high. If we see those numbers in relationship to each other, in other words, as ratios, then we can translate them into the sonic spectrum and have a corresponding set of ratios in sound. This is exactly how an ordinary scale works on any musical instrument. The sound of the scale depends on the relationship of adjacent tones to one another.

The question naturally arises at this point: If the ratios are actually those of light vibrations, how can they become sound?...

An important key to understanding how we can

actually hear high, fast, light vibrations is the Law of the Octave. This law states that any vibration of sound (or light) can be doubled or halved, and the same pitch (or light frequency) will result, but what changes is the octave of the sound (or radiation). A simple example: Orchestras tune to the concert pitch A, which is established at a frequency of 440 hertz (cycles per second). Playing the same note at 220 or 880 hertz results in a tone we immediately recognize as an "A," but it sounds either an octave lower or higher than the concert A as such. By taking a very rapid vibration of light and halving it many times (about 35 iterations), we can bring this vibration into the range of hearing.[47]

Astrophysicists measure radio waves near the speed of light from some quasar point sources that spread out their emissions.[48] I propose there are other vibrations, possibly more abstract than the tachyon, that interact with our large dimension and move faster than science can comprehend and measure. Both measurable and elusive vibrations may affect our DNA although we only hear certain lower octaves. We might receive and transmit all universal vibrations within our DNA template, at least partially.

Alexanjer's quotation states approximately 15 different wavelengths are measurable in each DNA nucleobase in unique variations. If these wavelengths in our current large dimension are generational expansions from a greater universe, then the number 15 could represent the dominant frequencies of a 15 large dimensional kathara grid. Before I address these large dimensions that are beyond M-theory, I will explore antimatter and parallel dimension potential.

There is antimatter within our galaxy, but it generally exists separately to matter because both will annihilate upon contact. An antiparticle contains the same mass as its particle twin, but its charge and other quantum properties are opposite.

In 1995, antihydrogen was isolated at CERN;[49] however, it

could not be measured and manipulated until about 17 years later when it was trapped in a very cold state. Initial measurements showed that antihydrogen is very similar to hydrogen. Physicist Mike Hayden declares, "It looks like an ordinary hydrogen atom. If there's a difference, everyone's betting it's going to be subtle."[50]

In general, physicists believe the following about antimatter after the Big Bang:

> As the universe expanded and cooled, almost every matter particle collided with an antimatter particle, and the two turned into two photons—gamma ray particles—in a process called annihilation, the opposite of pair production. But roughly a billionth of the matter particles survived, and it is those particles that now make the galaxies, stars, planets, and all living things on Earth, including our own.[51]

According to the Standard Model of particle physics, matter and antimatter were created in equal amounts at the Big Bang. In 2010, a smaller version of the LHC called the Tevatron gave eight years of data showing there was a one percent difference between the amount of matter and antimatter produced in the accelerator experiments, favoring matter.[52]

LHC researchers state that stable antimatter no longer exists in our discernible universe. Matter edged its asymmetrical counterpart out of creation after the Big Bang, but it is unknown how this happened.[53] Since matter is the dominant material in our galactic experience, its antimatter counterpart could fulfill its inherently equal potential if it exists as a lesser variable or coordinate in small, non-compactified dimensions or subspaces.

Rarely, a particle and its antiparticle can coexist when a force intervenes to create a pair bond that maintains quantum balance between the two. A physics forum member explains:

> Incidentally, the easiest way to induce pair production is to shoot a sufficiently high energy

photon (in the gamma frequency range) very close to a heavy atomic nucleus. As the gamma interacts with the dense electromagnetic field of the nucleus it will excite the electron field to produce an electron and a positron. The frequency of the gamma photon has to be high enough to provide the energy to create the rest mass of both particles (511 keV each) and to have enough left over to give them sufficient kinetic energy to escape their electromagnetic attraction.[54]

This description essentially deconstructs an intact nucleus to create an unstable antiparticle. The physics enthusiast continues to explain what happens when an electron and positron meet:

An electron wave & a positron wave are perfectly anti-symmetrical, so when they meet, they neutralize each other, "spilling" their energy content into the electromagnetic field. The two waves cancel each other out, each tweak & twist in the electron field being canceled out by its opposite partner.

Due to spin conservation, there are actually two (main) possibilities here: if the electron & positron have parallel spin, 3 (or a higher odd number) of photons will be emitted. If the electron & positron have antiparallel spins, then 2 (or a higher even number) of photons will be emitted. The reactions with fewer emitted photons are much more likely to occur; higher numbers are only rarely seen, unless huge energy levels are involved.[54]

The term *annihilation* implies complete destruction, but technically the electron and positron pair can release their energies toward another creation. If their energies continually recycle into new structures in our galaxy, then their potential for creation will eventually diminish into unstructured subatomic particles—space dust.

Any antimatter created by the Big Bang would likely pertain to observable galaxies, but if the Big Bang process mostly copied an earlier, less violent creational pattern, perhaps another galaxy was co-created along with a galaxy made of matter. While it is a stretch of the imagination to think there were enough stable antiparticles from the Big Bang to be transported to a large dimension nearby, I think there is a greater probability of an earlier "bang" that created a more stable and potentially eternal capacity for both antimatter and matter to exist at their full potential. This earlier process expands the cosmos to include entire systems, galaxies, and universes composed primarily or exclusively of this antimatter existing in a parallel position to matter.

In theoretical physics, a form of electromagnetic energy called the scalar wave exudes from a scalar field that is more basic than an electric or magnetic field. In mathematics and physics, a scalar field is a set of observable scalars (quantities with magnitude) at every point in space or a large dimension.

A team of custom electronics developers explains with input from Wikipedia, The Free Encyclopedia how scalar and typical electromagnetic energies differ:

> Temperature and pressure are scalar quantities and have no direction associated with them. Any electromagnetic phenomenon, however will always have a directional (vector) component associated with it. This is because a magnetic field is always a dipole (north and south), and is actually caused by the motion of charged particles or an electric current....
>
> Scalar electromagnetics (also known as scalar energy) is the background quantum mechanical fluctuations and associated zero-point energies (in contrast to "vector energies" which sums to zero).
>
> Scalar waves are hypothetical waves, which differ from the conventional electromagnetic transverse waves by one oscillation level parallel to the direction

of propagation, they thus have characteristics of longitudinal waves....

Scalar field theory suggests that scalar energy can move through space much like an electromagnetic wave. However, the operating principles are different. The regular expansion and contraction of a scalar bubble/void is like rythmicly [sic] splashing water on a pond. It sends out ripples through the general scalar field that can subtly affect the size and strength of distant scalar bubbles/voids.[55]

Magnets repulsing and attracting each other could respectively cause scalar bubbles and vacuums, which are not empty voids because they contain the lowest possible energy. The objective of a scalar communications antenna is to provide these magnets as two opposing electromagnetic coils to create large scalar bubbles and vacuums as a source of powerful energy. The problem with this idea is that the coils almost entirely cancel out each other's magnetic field, so the antenna does not emit a measurable electromagnetic field. The antenna merely heats up and is thus considered useless to standard electromagnetic theory.[55] Perhaps this is what scalar mechanics naturally do: produce less forceful and less dense energy-matter than what dominates our reality. Scalar fields and waves would differ from electromagnetic waves by penetrating "through materials that would normally slow or absorb electromagnetic waves."[55]

Quantum field theory describes a tachyon as a quantum of a scalar field. The MCEO takes this concept further and declares tachyons as fractal versions of earlier superluminal partiki takeyons,[56] which is a possibility when comparing tachyons that increase speed as their energy decreases to eternal takeyons that increase speed as their energy increases.

The Guardian Alliance describes pre-matter in the pre-galaxy cosmos: "*Partiki are the smallest units of energy in the cosmos* (one could find 800 billion billion Partiki units in an average 3-dimensional photon)" (p. 453).[57] Partiki units are omni-polar units that contain all polarities, states the ATI,TPE. They group into strings to form the kathara grid, the backbone of subsequent scalar and electromagnetic grids. According to the MCEO-GA and ATI,TPE, scalar waves appear to move from one place to another, but they are "standing" points of light strung together in sequences within the fabric of universal morphogenetic (form-holding) fields.[58]

Partiki manufacture "two intrinsic sub-units of crystalline morphogenetic substance [particum and partika] that serve as blueprints for rhythms of pulsation through which particles and anti-particles manifest" (p. 453).[57] Particum constitutes eternal matter, and partika (pronounced partikā) constitutes eternal antimatter vibrating faster than particum. Particum and partika clusters are keylons. To elaborate upon their prior introduction, keylons form specific geometric patterns as crystalline matrices of electromagnetic energy.[57]

The process of vibration between similar partiki units is called partiki phasing, wherein it never annihilates the polarities but continually circulates their energies according to different vibratory speeds in a process of gentle fission and fusion.[58] The partiki unit develops in stages to create the electrical abilities within the partika, clarifies the ATI,TPE, which the particum pulls in with magnetism as a collaborative merkaba vehicle that I will explain near the end of this section.

Krysthal spirals have naturally taken partika and particum matter in different directions by carrying the kathara grid template to new spaces for their development. The Milky Way galaxy is based upon a kathara grid with a particum template, meaning that partika-based antimatter exists out of our view with an entirely separate, galactic kathara grid. The electromagnetic antimatter based on the partika unit is different from the antimatter within our particum-based universe.

The MCEO-GA and ATI,TPE state that the Milky Way has a parallel 3rd dimension as well as a parallel kathara grid comprised of partika-based antimatter. Their parameters involve slightly different forces and angular momenta of particle spin positioned outside of our dimensional experience and view, while existing in relative proximity to us. I argue that a parallel creation has an individual consciousness, makeup, and reality that are independent of ours, but it contains a familial energetic connection.

Although we cannot see parallel antimatter, we can partially see higher, large dimensions in our Milky Way because they are made of somewhat similar particum-based units to which we are fundamentally keyed. More precisely, our view of these higher dimensions is skewed because of light refraction.

The MCEO-GA have stated that our complete time matrix contains 15 large dimensions; however, they have only revealed star and planetary locations with a 12 large dimensional, kathara grid formation, causing some confusion as to what constitutes a time matrix.[20,58] The MCEO-GA and ATI,TPE agree that the galactic structure is built with the kathara grid formation, but they differ in terms of the kathara grid size. The ATI,TPE reveals that the complete galactic kathara grid contains 15 large dimensions.

To build upon the prior definition of a large dimension, each one provides a density carrying a dominant frequency of energy among a blend of frequency hues. The terms *large dimension* and *density* can be interchangeable in this context. In addition to our three-dimensional experience, large dimensions may also group into threes according to their dominant energy-matter. Three consecutive, large dimensions would constitute one harmonic universe (HU) of overlaid reality fields and waves of matter. We live in HU-1, which currently has the most extended and dense matter. (Note: from this point forward, *dimension* will equate to *large dimension* unless clarified otherwise.)

In HU-1, the Earth and its organic inhabitants are predominantly made of carbon. The carbon atom has six electrons with four of them in its outer energy shell.

HU-2 has a faster harmonic oscillation of light in dimensions 4, 5, and 6 where life forms are silicon-carbon based.[58] The silicon atom has 14 electrons with four of them in its outer energy shell. When multiple carbon or silicon atoms come together, their outer four electrons called valence electrons can bond to form a crystal.[59] As crystalline silicon-carbon, the HU-2 level of the kathara grid will comprise a slightly different elemental configuration in order to absorb and produce more light for the increased electrons. Compared to HU-2, the HU-1 particle pulsation of time is faster, and the vibration of matter is slower.

Since HU-2 has its own celestial and planetary bodies, and they originated before the Earth in the stair-step creation model down toward HU-1, it is probable that an HU-2 counterpart to Earth exists and has imparted some of its energy to Earth and its inhabitants. According to the MCEO-GA and ATI,TPE, this counterpart to Earth does exist and is called Tara. These sources also state that silicon-based HU-3—dimensions 7, 8, and 9— contains a less dense counterpart to Tara called Gaia.

The names Tara and Gaia are well-known. Tara is a Hindu goddess, and she is also a female Buddha.[60] When combined with the Latin word *Terra* for Earth or land, New Age religions say "Mother Earth." New Age believers additionally call our Earth *Gaia*, or they equate the "spiritual" level of Gaia to our Earth. Gaia was the Earth goddess of Greek mythology.[61] Regarding the Earth as a woman may appear to revere the woman for her life-giving, reproductive ability, but really, this association continues to lock women into the sexually reproductive role. In addition, these supposedly progressive belief systems continue to put men above women in their characterization of Father Heaven over Mother Earth. Male and female sexual identities do not represent any planet, star, or "heavenly" space; these are limited perceptions and distortions to their true natures.

The HU-4 level of dimensions 10, 11, and 12 primarily contains hydroplasmic liquid light. The GA designates entities in this harmonic universe as avatars.[20] Hinduism defines an avatar as a deity who descends to Earth either as a manifestation of its otherworldly, higher dimensional nature or an incarnation into human or animal form. The term *avatar* is a title given by entity groups that may or may not apply to HU-4, and it may be irrelevant to the actual entities in HU-4.

As Figure 3 illustrates, the Milky Way kathara grid ends at the 11th dimension (more precisely, the 11.5 dimension). The proper HU-4 counterpart to Earth is not popularly known because this harmonic universe is severely compromised in the Milky Way galaxy. The ATI,TPE reveals that both 10th and 11th dimensions are compacted with an HU-3 silicon base, which gives our galaxy five densities of silicon material. The book *Eternal Humans and the Finite Gods*, third edition, provides a plausible history of how god-playing entities have manipulated and damaged our galaxy.[17] M-theory can potentially support Milky Way creation with 11 dimensions and measures of compaction in both large and small scales.

I am revealing with information from the All That Is, The Pure Essence that the MCEO-GA's description of HU-5 is erroneous due to their shortened, 12-dimensional kathara grid. Their presentation of creation requires HU-5 to exist in an adjoining kathara grid, which is what the MCEO-GA assert; however, they claim that the 13th dimension exists at the top of this grid as though it carries more powerful frequency than dimensions 14 and 15.

I argue that HU-5 exists as the top harmonic universe in a 15-dimensional kathara grid with dimension 15 logically at the apex. The 15-point kathara grid is the Krysthal kathara grid. Contrary to what the MCEO-GA teach, the ATI,TPE states that HU-5 does not consist of primal light fields, nor does it contain Breneau Orders or Rishi entities.[20] I have not gained an answer about what type of entities exist in HU-5 because they prefer to be unknown for security reasons. The 15th dimension contains the fastest dimensional frequency from where particle spin and vibration progressively slow into the lower dimensions. All 15 dimensions of a galactic kathara grid constitute one time matrix.

The ATI,TPE states that Breneau Order or Rishi entities exist in the HU-1 of the preceding domain of creation called the Ecka. Rishi are ante-matter constructs of thermoplasmic radiation, states the GA; if this definition is correct, then this "matter" would largely constitute the life forms in those particular Ecka levels.[20] The Ecka domain contains a 15-dimensional kathara grid foundation similar to the galactic kathara grid but with less dense material. The galactic domain is called the Veca.

A domain incorporates various levels, which we experience as large dimensions in our galaxy, and each domain can have numerous realms. The Milky Way galaxy is part of a Veca realm within the greater Veca domain.

A structurally healthy galaxy has one or more stargates in each large dimension, depending upon the creational matter and size of the individual large dimension, states the ATI,TPE. A galactic stargate is a natural gateway formation extending outward and entering other locations within and beyond the galaxy. It is formed during the beginning stage of a large dimension, and it becomes incorporated into a star or planet that is also formed at that position. Most galactic stargates exist on a lateral axis within a celestial body, producing horizontal then angular passages through the galaxy, informs the ATI,TPE.

Death-based science and technology can siphon natural galactic material to create other galaxies of lesser structural integrity, as Figure 3 shows in relation to the Milky Way. Distortions can be prevalent in the Veca domain, but they can also exist in the Ecka domain due to similar tactics there.

Figure 3 compares a simplified drawing of the Krysthal kathara grid with three progressively fragmented and distorted galactic copies in our immediate realm. The ATI,TPE reveals the name of our Krysthal galaxy: AquaLaSha (pronounced AquaLaShā´). Each diagram is a basic approximation of kathara grid relationships and sizes. Background, non-dimensionalized components are implied in their construction. As shown, the Milky Way grid is wider than the AquaLaSha and Galaxy-2 grids due to having additional components and "junk," and the highly distorted, "phantom" extension off of the Milky Way grid is significantly thinner due to severe compaction.

Energetic bridge-like formations have been constructed to extend between galaxies in order to maintain Krysthal connection, such as the important bridge system between Galaxy-2 and Milky

Figure 3. Simplified Krysthal and Unnatural Galactic
Kathara Grids in our Realm

AQUALASHA AND GALAXY-2 GALAXIES

11-DIMENSIONAL MILKY WAY GALAXY

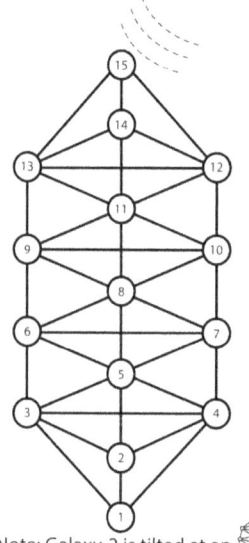

* Note: Galaxy-2 is tilted at an 11.75 degree arc to the right

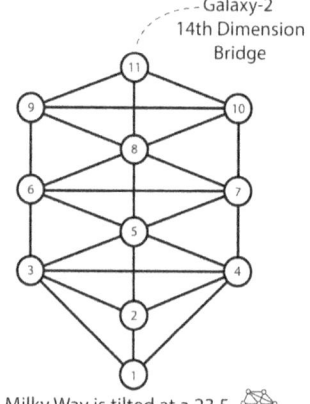

Galaxy-2
14th Dimension
Bridge

Milky Way is tilted at a 23.5 degree arc to the right

PHANTOM MILKY WAY GALAXY
Tree of Artificial Life

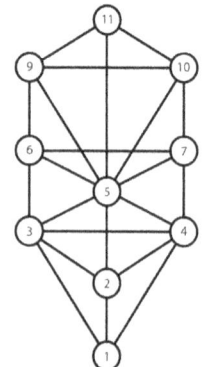

Phantom Milky Way is tilted at a
25 degree arc to the right.
* The Tree of Artificial Life tilt
varies per galaxy it distorts.

Way. The ATI,TPE reveals that a one-way bridge, which I name Bridge-A, was built in Galaxy-2's 14th dimension to provide a natural energy flow and shielded pathway to the Milky Way's 11th and highest dimension by way of a detour through Galaxy-2's 12th dimension. Bridge-A bypasses the longstanding damage of the 12th dimensional stargate caused by an ancient dominion war; only recently in early 2012 has most of the stargate been repaired, and it is in the process of establishing secure connections to other dimensional stargates, states the ATI,TPE. A second bridge, which I name Bridge-B, was constructed between our 11th dimension and Galaxy-2's 12th dimension to restore our access to the highest dimensions of a complete kathara grid. "'Bridge-B' is a holding passage [elaborated as a short, precise containment area] whereby life forms prepare and further transfigure their essence to accommodate their transition to the 12th dimension in Galaxy-2," explains the ATI,TPE. Bridge-B carries the necessary frequencies and codes to enable our ascension to the 12th dimension, and it prevents dangerous predators and destructive entities from entering.

In AquaLaSha's and Galaxy-2's 15-dimensional kathara grids, the positions of dimensions 12 and 13 do not follow the same pattern as the lower dimensions. The following response by the ATI,TPE addresses my questions about this simple but non-uniform kathara grid diagram.

> There is a gap of four dimensions between dimensions 9 and 13 while the other gaps have two or three dimensions in AquaLaSha's kathara grid because the spaces widen in the ethereal stairway as it progresses in and beyond that 9th dimensional level. These spaces contain less compaction of galactic "junk" that has collected in the lower levels and dimensions, causing more expansion and freedom of elasticity of spatial and aerial elements.
>
> The lone four dimensional space is not a four-directional, flat plane. The space of four dimensions

has more volume and is not equal to the two dimensional space on the opposite side of that grid.

To clarify, a completely uncompromised Krysthal kathara grid does not have "junk." It has residual remnants from natural components that are referred to as particulates and particle stream formations. The Krysthal Ecka kathara grid expansion contains more ethereal gas formations and substances that are also not referred to as "junk."

In the 2016 edition of this cosmology book, I created Figure 3's 15-dimensional kathara grid based on existing kathara grid presentations along with my intuition and new information provided by the All That Is, The Pure Essence. I initially thought it was sufficient to explain the new information without providing a significantly different kathara grid; however, my awareness grew to determine the existing kathara grids unacceptable for the following reasons.

The conventional kathara grid diagrams show vertical and horizontal energetic flows between large dimensions; however, they are actually lateral by an angular pathway, which is not necessarily diagonal. The ATI,TPE explains, "The language 'angular' is more accurate for kathara grids, for their lateral pathways are not consistently diagonal in direction but can be curved differently with small vertical or horizontal flows."

The encircled numbers depict large dimensions and their respective galactic stargates somehow existing in the same place as a specific star or planet. This presentation under the Law of One paradigm misrepresents evident reality. Therefore, the circle should be removed around each distinct large dimension to allow room for expansion.

The kathara grid is not completely rigid, although it does have a structural pattern. While each large dimension has a placement upon the grid, its inherent creativity is not contained. There are also different spatial and energetic configurations in higher or earlier densities, as I have already established. This means that a kathara grid is not created by a mathematical equation seeking uniformity. The structural pattern of a kathara grid contains a measure of randomness.

AquaLaSha has an uncommon galactic kathara grid, but it is still Krysthal and eternal. The creation process of AquaLaSha's grid and its top four dimensions naturally flowed in another direction than how the majority of Krysthal galaxy kathara grids form. Additionally, a Krysthal kathara grid in our Veca domain is somewhat different from a Krysthal kathara grid in the earlier Ecka domain, confirms the ATI,TPE.

For these reasons, I decided to create a more accurate diagram of Krysthal galaxy kathara grids comparing AquaLaSha and the common kathara grid in Figure 4. I applied the information given by the All That Is, The Pure Essence and asked for further details. Although these kathara grids only approximate foundational patterns for a Krysthal galaxy, they can assist our perception of spacetime organization.

In these more accurate kathara grids, the dashed lines represent the lateral energetic connections between large dimensions. The gray shaded background within and including the kathara grid outline represents the minimal, pre-matter structural template of potential space expansion wherein nothing has yet been created, and there is no light. There are no circles around the numbered dimensions, which normally do not expand beyond the shaded background. The kathara grid also extends further outward in dimensions 12 through 15.

The 15th dimensional top of AquaLaSha's kathara grid naturally connects to the 1st dimension of the earlier Ecka realm, and it does so with an arc-like energetic bridge that curves to the right (in Figures 3 and 4), states the ATI,TPE. The Ecka has extended this connection to Galaxy-2's 15th dimension. This curved connection is not part of the Krysthal spiral.

The ATI,TPE clarifies that the Krysthal spiral stream extends from the core of an Ecka kathara grid to the core essence of AquaLaSha, which first started the creation process of AquaLaSha. This stream also extends to the core essence of Galaxy-2, providing a viable foundation that bypasses the distorted aspect of created matter. The kathara grid's core is located within the central 8th dimension, and it holds the morphogenetic field for the entire galaxy. The core connection of successive kathara grid formations is integral to eternal energy flow. The Milky Way's 8th dimensional core has become compromised beyond minimal Krysthal spiral connection, so our galaxy needs the connection to Galaxy-2 in order to receive the complete 15 dimensional frequencies.

Figure 4. Corrected Krysthal Galaxy Kathara Grids

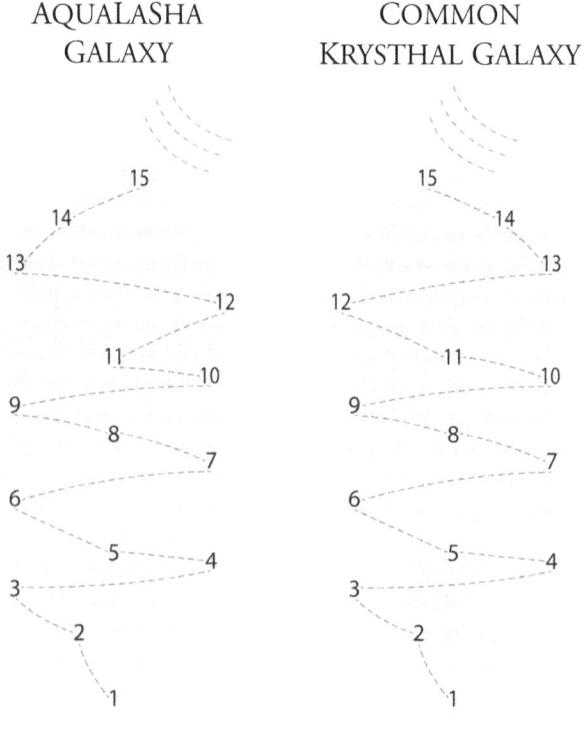

"Galaxy-2 was created as an offshoot to the right of the same Ecka kathara grid core [that also connects to AquaLaSha], which bridges it to AquaLaSha's natural grid," elaborates the ATI,TPE. Because the Ecka is a different domain with distinct energy-matter, its spatial position is mainly abstract to our reality. Therefore, it is feasible for Figure 3's kathara grids to connect in a direction to the right toward increased energetic integrity, which also reaches toward the bottom or edge of the Ecka, while at least some other Veca kathara grids connect in a fundamental manner to the Ecka in a direction to the left.

From our Veca position, the Ecka's HU-1 dimensions can be perceived as three arc-like waves, as shown in Figures 3 and 4. The Ecka has its own process of stair-step creation with 15-dimensional kathara grids similar to ones in the Veca, but their large dimensions have less boundary than our experience. When correcting the shortened and doubled kathara grid model that the MCEO-GA place within the Veca, the Ecka dimensions 1, 2, and 3 would constitute the "Primal Light Fields," and they would be preceded by the "Primal Sound Fields" (p. 299),[20] which appropriately would be assigned as Ecka dimensions 4, 5, and 6.

The Yanas entities exist within the Ecka's HU-1 and HU-2; collectively, they are called Eieyani, as confirmed by the ATI,TPE. From our earthly perspective, Eieyani look like "geometric shapes made of living light," states the GA (p. 159).[20] The MCEO entity group originated as Eieyani before they decided to get involved with creation modalities in the Veca where the GA resides.[17] Various god-playing entities in higher dimensions say they are the Word and Light of our more complex density and life, when light and sound exist throughout the Ecka-Veca in respectively unique and valid ways regardless of any minority claiming to monopolize them.

The MCEO-GA creational model of spirit-science is one of replication and proportion, which significantly minimizes variation. Their 12-point kathara grid creates a central square containing dimensions 3 through 10. This square is able to house four kathara grids when creation expands within it.

The MCEO-GA version of a Veca encapsulates within a sphere four kathara grids within the central square of a larger

Ecka kathara grid. The four kathara grids are a matter-based pair and a parallel antimatter-based pair, both containing 15 total dimensions and what the MCEO-GA call the Energy Matrix as the "primal" sound fields, and the 12-point grids connect at a 90 degree angle. The Ecka's kathara grid is rotated 45 degrees to the right, which conforms to Figure 2's 12-Point Kathara Grid Spiral.[21] This pattern of grids within spheres supposedly leads to the beginning of creation.[62] However, the MCEO-GA's origin of creation is limited to their chosen star or spherical realm, which would exclude prior levels and other realms outside their creational model.

I asked the ATI,TPE how many galactic kathara grids are in our natural Veca realm, excluding distorted copies. It states that our Krysthal Veca realm is a 15-galaxy system that was created by our Ecka's material expansion process. Fourteen galaxies come in matter and parallel antimatter pairs while one 15-dimensional grid is parallel antimatter without a counterpart composed of matter. Although this is our realm's pattern, it is not necessarily the pattern for another natural Veca realm. The ATI,TPE also states there is "no particular angle" at which kathara grid pairs and an odd remainder must relate to one another.

Although the MCEO-GA's base-12 mathematics has its flaws, it carries more structural integrity than base-10 mathematics. The last kathara grid representation in Figure 3, the Tree of Artificial Life, is the most distorted one, yet it is heralded in New Age and Jewish Kabbalah "sacred science" teachings as their Tree of Life, otherwise known as the 10 Sefirot (or Sephiroth).

The Tree of Artificial Life removes the critical 12th and 8th dimensions, essentially creating a 10-dimensional template while preserving the 11th dimension, just barely. It cuts off nearly all incoming natural energy and is the next step toward creating a fully finite and fully phantom grid. Eradicating the kathara grid core's form-holding field, as well as our form, is the goal of death science.

When something has a phantom status, its portion of the kathara grid may still have a link to the original grid by means of intermediate energetic connections and highly aware beings keeping it alive. Similar to the Fibonacci sequence's increasingly detached stages from their source, there are several stages of reconfiguration from eternal matter to implosion and eradication.[17] The phantom component is vampiric, meaning that semi-phantom and phantom entities must assimilate energy quanta from other systems to artificially extend their lives as possibly immortal but non-eternal beings. They have the potential to gain their original blueprint's structural integrity; however, when fully phantom matter is compromised and compacted past the point of repair, any prior connection to eternal energy is severed, and implosion to space dust is inevitable.

The All That Is, The Pure Essence consistently confirms that matter was initially pulled off key locations in AquaLaSha's galaxy by interfering entities to eventually create the Milky Way and then Phantom Milky Way with increasingly altered, phantom energy-matter. The Milky Way is an incomplete galaxy whose parts have varying measures of semi-phantom to phantom status. Our galaxy is in a precarious position that can be ameliorated by shifting our focus and insights toward true non-death energy, not the artificial life energy of vampiric immortality taught by self-serving religious entities.

What concerns us from our position in the Milky Way is how we can attempt to leave our density and potentially become Veca "ascended masters." To clarify this term, Hinduism and Buddhism, along with other religions such as Christianity, usually believe that we become Ascended Masters after we die and reach a vague heaven that actually exists within HU-1 or HU-2; then, we may return to Earth with full awareness. This belief assumes other dimensional levels give us exceptional enlightenment when they are not significantly different from ours, and the reincarnation process somehow does not diminish that awareness. A New Age variation of this belief says we can master a cycle of numerous reincarnations, which assumes a power greater than the severe fractalization of seven continuous reincarnations, and this brings enlightenment and immortality as an Ascended Master "to help the rest of humankind."[63] Realistically, an ascended master is merely any galactic entity who has mastered the galactic ascension process by transforming and traveling up the complete 15 dimensions to the Ecka, and the entity can proceed from there as it chooses.

Our ascension through the dimensions is taught in New Age spirit-science as a vertical process, but it is really an inward process to the higher dimension. I deduce that entity groups promote vertical, not lateral, ascension to humans because our body has a central, mainly vertical axis of energetic structures such as chakras that manipulative entities desire to control. In fact, the All That Is, The Pure Essence rejects the "vertical" description for ascension.

Natural ascension involves a half dimensional step downward to enter into the deep physical core of the planet, to then laterally by an angular pathway emerge one full dimension higher. Ascension requires the DNA and bodily components to fold into themselves and then laterally, minimally implode in the course of mildly to moderately reconfiguring or transfiguring to a lighter density. When the ascension process is natural, it transfigures the body with increased Krysthal energy-matter, and it sloughs off and leaves behind as residue the density's phantom matter and implants.

The ATI,TPE explains the bodily ascension process:

> The transfigured "body" goes through the expulsion process, laterally stepping out the dimensions to stargate-12 through the Galaxy-2 bridge and then to the Ecka. The term "laterally" for the body essence ascension pathway refers to a side-stepping process which also includes a diagonal stepping outward through dimensions to the destination intended.

According to the kathara grid formation and its embattled distortions in the Milky Way and Galaxy-2, there exists significant distance between Earth, Tara, Gaia, and Aramatena, which is the more natural 12th dimensional counterpart to Gaia in Galaxy-2. The MCEO has stated that Tara is the star named Alcyone in the Pleiades,[64] but the ATI,TPE reveals that Tara was partially created by a portion of Galaxy-2's Alcyone in the much higher 13th dimension. In addition, the Pleiadian Alcyone is not a single star; it is a multiple star cluster. Our historical connection with Tara has prompted the plethora of Pleiadian god-group entities, who have long influenced our beliefs through prophets and "saviors" including Jesus and Buddha, to amalgamate the original Alcyone name with Tara and their highly regarded star cluster in the Pleiades.[17] The ATI,TPE states that Tara is actually a planet in the northeastern tip of the Pleiades star system.

Many stars and planets came into existence by both an organized and disorganized pattern. When the galactic components and kathara grids became subjected to artificial technologies by interfering otherworldly entities, they caused versions of Earth, Tara, Gaia, and Aramatena to alter their pole alignments, angular momenta of particle spin, and positions. This problem and the curved reality of spacetime can cause significant perceptual and actual spatial differences toward similar planets and stars that are divided by time and density.

The process of ascension can have us transcend these physical distances when the specified planets are properly energetically and spatially aligned by their centrally located vertical and horizontal axes, and their merkaba vehicles and fields. When ascension may happen in the Milky Way, the body can shift its angular momentum and density of atoms toward its proper destination, which could be Tara's silicon-carbon existence unless it is obstructed. Currently, the closest and most viable pathway for our natural ascension is directed from our current 2.5 large dimensional position to the 3rd large dimension where the preserved, original aspect of the Earth, called Amenti Earth, remains with a sufficient portion of its energy-matter. As is the theme of death science creation, a significant amount of energy-matter was pulled from Amenti Earth and further distorted with various electromagnetic netting, including a "poison apple" torus configuration, and other technologies to form lesser copies, resulting in a displaced and weakened Earth.[17]

When the opportunity may arise for our ascension to remove death science mechanics, we would not travel according to our observational distance; we would merely shift aside and experience a transfiguration process that both matches and blends the new energy-matter of the lighter density with what already exists within our DNA template. We gained a carbon base on Earth that has the potential to gradually transform to Tara's slightly different configuration that then becomes a fully silicon base on Gaia. Then, through Galaxy-2's Bridge-B (or eventually the 12th dimensional stargate), the silicon base can transform to a primarily crystalline liquid-light, hydroplasmic state on the 12th dimensional star Aramatena. From there, we can transfigure to a lighter base than hydroplasma in the 13th to 15th dimensions and enter the Ecka. These planets and star originally contained the most resonant energies and cycles, but we are not limited to their destinations.

The lateral ascension process through previously created large dimensions is a natural pathway for our ascension, states the ATI,TPE. There is also another natural ascension pathway that occurs directly between corresponding dimensions of the Ecka and Veca domains because of the Tri-Veca subatomic code built into Krysthal creation. This code causes Ecka stars and planets to create similar Veca structures (that continue to be formed by galactic events) in or near the same dimensional position of their respective kathara grids. The Tri-Veca ascension pathway is protected and thus preferable due to its fully Krysthal nature.

In Figure 5, the Tri-Veca code has an upper partiki unit that splits off a measure of itself to create partika and particum versions in the process of expansion into a new domain. From our perspective, the original Ecka counterpart to Amenti Earth can be omni-polar because its energy-matter combines both antiparticle and particle divisions, but it is also a polarized, particum-based planet or star that was mostly created by an earlier Eckasha krystar. The Tri-Veca pathway of creation and resultant ascension provides an expeditious flow of energy akin to a partiki unit in its phasing cycle. Although this is an eternal code, it is also used in partially distorted realms where sufficient energetic integrity remains. For instance, a portion of our familial Ecka kathara grid was harnessed in a similar way as AquaLaSha to create a partial, secondary kathara grid containing a distorted version of at least the first two harmonic universes. The entities in this kathara grid—some with the flow of natural energy with pure intent and others with an agenda—utilized the Tri-Veca and Bi-Veca codes to largely create Amenti Earth.

Figure 5. Tri-Veca and Vesica Piscis Codes

BI-VECA
TRI-VECA   "VESICA PISCIS"

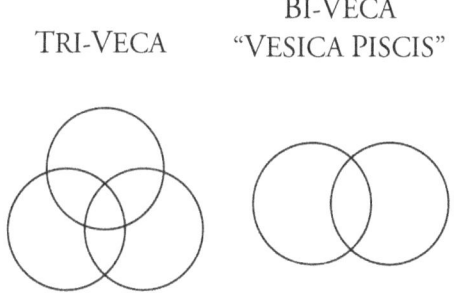

The Bi-Veca or vesica piscis code is prominent in "sacred" spirit-science teachings in the New Age community, and it is incorporated into architectural designs of churches and cities. The vesica piscis is a technology that severs the partiki at the pre-subatomic level and creates polarized, "dead" light. The death science components of our Earth and Milky Way galaxy were formed by this distortion.

On a related note, there is speculation about a literal Inner Earth in which our Earth is hollow, and it contains another world of inhabitants. Some people claim there are holes at the north and south poles providing access to the hollow Earth. Admiral Richard E. Byrd noted the "Great Unknown" he encountered when leading United States military expeditions in 1947 and 1956 over the North and South Pole, respectively.[65] He traveled excessive miles beyond both poles, reporting a large land uncharacteristic of the icy regions with abundant greenery.

Instead of a hollow Earth that would invalidate the geological model of the Earth's core and mantle, this unique region could be an interdimensional zone near the Earth's core, and any polar entrance could be a noticeable, energetic vortex that intersects the two worlds. God-group entities have taught the New Age community about this otherworldly region they call Inner Earth. The term "Inner Earth" is also used by Law of One god-group entities to signify Earth's natural Ecka counterpart containing the Tri-Veca code. To buffer the energetic transition between Amenti Earth and the Ecka "Inner Earth," there is a 3.5 large dimensional modulation zone at the core of the Sun.[57] This term is another example of how different locations are combined and applied to our reality when they are difficult to verify in our density, so their stories need further investigation.

The order of natural domains that extend beyond and before the Ecka are as follows: Eckasha, Eckasha-A, Eckasha-I,[62] Eckasha-Aah, Eckasha-Ah, and A (pronounced with a long Ā), states the ATI,TPE. An Eckasha level contains varying stages of pre-partiki, etheric essences in the early domains to an increasing plasmic and silicate structure in the later domains.

The ATI,TPE reveals that the Eckasha-A domain started the 15-dimensional kathara grid formation, and the much earlier Eckasha-Ah domain started the Krysthal sequence for creational expansion. The Fibonacci sequence started later in the Ecka domain.

Initially, Ashayana only taught about levels up to the Yunasai, which are a group of Yunasum partiki, because her entity allies in the MCEO-GA said the Yunasai are the "Central Point of All Union," even calling them "God" and "Source" (p. xlix).[20] MCEO-GA members have perceived the Yunasai or singular Yunasum as their point of creational reference because subsequent creations contain an extended aspect of its light-body structure.

However, the Yunasum actually contains 20,736 partiki units; earlier partiki units created the Yunasum.[66] The ATI,TPE states that the first partiki element originated in the Eckasha-Aah domain. Partiki units act as building blocks, and their grouping can potentially create a small krystar, not a domain.

Direct communication with Yunasai consciousness via DNA keylon "fire letter" transmission, which I explain in the next section, reveals their perspective to my mother who has this ability: "We are krystars, and a Yunasum is a singular krystar which there are more than millions." The ATI,TPE elaborates with my additional questioning that the Yunasai exist in a level throughout the Eckasha-Aah domain.

The Yunasai exist beyond the Starfire flow of energy, which is a far-reaching process of ascension or partiki phasing that utilizes the Tri-Veca code, and circulates and regenerates energies between the Veca domain and what the MCEO call the Edonic worlds.[66] The Starfire process can also be expedited apart from the natural cycle by an in-built immune response within Krysthal creation to correct significant damage inflicted upon it; this damage has thus far occurred only in Veca and Ecka domains. The Edonic worlds exist in the Eckasha-I domain after the Yunasai, state the Yunasai and ATI,TPE. To be more precise, the original Edon partiki unit precedes the Yunasum unit, but there are slightly different reproductions of every partiki unit in each latter eternal realm, meaning that the Eckasha-I has both Edon and Yunasum generational formations.

Because the Yunasai are just beyond the Starfire ascension process, I deduce that many MCEO-GA entities consider them as a collective "God" beyond what they know as creation. If the MCEO-GA directly communicate with Yunasai as my mother has done, then they should discover that the Yunasai do not consider themselves to be a God or Source of all creation. On the other hand, perhaps members of the MCEO-GA did communicate with some Yunasai, and they received a response without diligent elaboration that they fit into a narrow perspective.

The ATI,TPE adds, "There is more beyond the Yunasai. The

MCEO, GA, and Yunasai do not have full knowledge. What is only known by them is revealed." Several years after Ashayana presented the GA's creational perspective beginning with the Yunasai, a more expansive view has opened up to her to show there are indeed earlier levels before them.

In the partiki-based light-body domains, entities can travel and communicate naturally or expeditiously with the use of merkaba constructs or vehicles. Personal merkaba vehicles may look similar to orbs when traveling, but their inherent energetic integrity can allow an entire being to naturally travel to eternal realms. "The personal merkaba vehicle structure," states the ATI,TPE, "takes the shape of a partially illuminated, spiraling plasmic vehicle of transport. It is different from an orb that is utilized in the post death process." It continues, "Orbs are an artificial energetic construction. They are similar to capsules that are released once they reach a phantom destination, thus allowing the more pure essence to return to its original home beyond the phantom created matter."

The top electric spiral of the merkaba spins clockwise (CW) while the bottom magnetic spiral spins counterclockwise (CCW). When the merkaba spiral sets are energetically charged, they create two merkaba fields. The merkaba structure looks like a three-dimensional star tetrahedron wherein the magnetic spiral is inverted under and mostly through the upright electrical spiral, both forming three-sided pyramids that effectively circulate energy.[57]

The ATI,TPE and MCEO-GA state that every living thing in a 15-dimensional time matrix has two personal merkaba fields as a natural part of one's anatomy. In addition, each harmonic universe and large dimension has two big merkaba fields that surround the smaller ones of their respective regions and inhabitants.[57] These big merkaba fields are created by the interaction of the galactic stargates and time. The ATI,TPE adds that higher dimensional merkaba fields do not encompass lower dimensional merkaba fields; however, a galaxy has a large merkaba that technically encompasses the entire galaxy.

Stargates as interdimensional openings naturally allow energy circulation and travel through the galaxy. While they produce horizontal then angular passages, time portals produce horizontal then diagonal (increased angle) passages that direct energies within the same harmonic universe. When they open up at the same time, a Seed Crystal Seal is unlocked for the dimensional transmutation.

The Guardian Alliance explains:

> Star Gates and Time Portals exist as Black and White Hole Pairs that are connected at the center point by a scalar-wave frequency Seed Crystal Seal. When the center Seed Crystal Seal releases[,] the Star Gate activates and the pair of counter-rotating electromagnetic spirals merge to form an interconnected Merkaba Field [fields], which allows the instantaneous passage between various spacetime coordinates through shift of atomic Angular Rotation of Particle Spin (p. 506).[57]

> All Suns have sets of black and white holes at their core; they operate as portals through which energy can pass through dimensional fields (p. 5).[57]

The Melchizedek Cloister Emerald Order explains how a merkaba can become vampiric:

> "Twisted" Merkaba Vortex mechanics implement unnatural distortions of the spin-speed and spin-direction of Merkabic Vortex sets, to create a particle/anti-particle harness field within which energy and atoms can be trapped. Once trapped within the inorganic Merkaba Field, the harnessed energy quanta can then artificially sustain a prolonged longevity of form, and achieve limited local interdimensional transport, as long as the inorganic Merkaba Harness around it can "feed"/drain energy from organically

living energy fields.[67]

There is competing information about merkaba mechanics given by a galaxy-based, Melchizedek-channeled entity group, headed by the entity Thoth, to a man named Drunvalo Melchizedek who is popular in the New Age movement and who supports the Tree of Artificial Life grid. He uses a reversed spin ratio of 34:21 that is 34 times CCW over 21 times CW, and he states that the vehicle around our bodies forms a disk of 55 feet in diameter.[68] When we refer back to Figure 1, we see these numbers representing the Fibonacci progression toward the golden ratio. The spin has a speed, and it is a numerical expression.

If the Earth's merkaba has a spin-speed just past 55 rotations, its internal energetic rods that influence the Earth's core to its crust would likely start a destructive pole shift, which was intended by gods such as Thoth, otherwise known as the Mayan god Quetzalcoatl, for the year 2012. The pole shift involves shifting the merkaba's axis and sometimes reversing the geomagnetic field, not a phenomenon flipping the Earth upside-down, for example. When the spin-speed of the merkaba's axis is altered, the shift changes the speed of one or both electrical and magnetic components, or reverses their directions. Geographical pole shifts, on the other hand, slowly occur on Earth with negligible effects.

To accelerate the Earth's merkaba to 144 rotations, each per trillionth of a billionth of a nanosecond, our Sun's larger merkaba fields would involve a manipulated blend with a phantom solar merkaba's fields to create a one-way feeding spiral, which consumes, destroys, and transfers the solar system energy-matter via an intrusive spacetime rip and wormhole system to a phantom galaxy where it becomes a large, external merkaba. The external "Death Star" merkaba would then decelerate to allow the trapped, raw energy to form into a mutated, finite-life copy of the two original systems, state the MCEO-GA.[67]

The number 144 is given great significance in religious teachings by god-group entities. For example, Aristotle and Plato, who were both heavily influenced by Thoth, taught that a significant change in the cities would occur every 144 years, and a significant change to humanity would happen every 1,728 years; both numbers are multiples of base-12 mathematics.[69] The Bible's Book of Revelation states that 144,000 Jews will be saved from apocalyptic judgement and given a New Jerusalem, whose city wall measures 144 cubits (Revelation 7:4-8; 21:17). The Law of One states that we have 144 versions of ourselves living simultaneously in HU-1 of the Milky Way (possibly also including the Parallel Milky Way, depending upon the interpretation), because the supposed oversoul aspect of oneself in HU-3—stated as a single aspect regardless of whether this formula includes both Milky Way matter and Parallel Milky Way antimatter—divided itself into 12 entities in the HU-2, who then repeated the base-12 pattern for the HU-1.[57,70] When expanding this formula to HU-4, it supposedly creates 1,728 simultaneous selves in HU-1.

The natural merkaba spin is CW electrical over CCW magnetic. The MCEO entity group and ATI,TPE state that the organic 3rd dimensional merkaba has the spin-speed ratio of 33⅓ CW electrical over 11⅔ CCW magnetic,[21] meaning that this merkaba is in the AquaLaSha galaxy. The ATI,TPE reveals that Amenti Earth has the merkaba ratio of 32⅓ CW electrical over 10⅔ CCW magnetic, just less than its Krysthal potential.

A Krysthal system can have natural, small black holes, but it does not have supermassive black holes or collapsed stars that could create a black hole. Physicists are increasingly agreeing that black holes act as wormholes. Wormholes spiral or bridge energy between dimensions and other universes of matter or parallel antimatter.

The Hubble telescope has merely glimpsed into pictures of an unfathomable number of galaxies surrounding the Milky Way. I have wondered if the observable galaxies and nebulas are comprised only of similar semi-phantom matter to ours. Upon my probing with the awareness of the All That Is, The Pure Essence, it states that fully phantom galaxies, and conversely fully eternal galaxies, are not visible to us. Their energy-matter have different compositions and angular momenta of particle spin than our particular matter. Accordingly, our semi-phantom matter is different from significantly more or less distorted, semi-phantom matter, which is why we cannot see Galaxy-2 outside of potential extrasensory, interdimensional ability.

The M31 Andromeda galaxy is visible to humans because it is semi-phantom like the Milky Way, and it was eventually created from remnants of AquaLaSha as was the Milky Way, states the ATI,TPE. M31 Andromeda is positioned approximately 2.5 million light years away from our Earth, but the gap is slowly diminishing between Andromeda and our galaxy. When viewing M31 Andromeda through the Hubble telescope, we can see it has a supermassive black hole in its center, but unlike the Milky Way, it has another structure connected to it.

I asked the ATI,TPE for clarification about M31 Andromeda. "There may be a double structure nucleus in Andromeda. Scientists say Andromeda either 'ate' another smaller galaxy, or the structure is merely a lopsided disk of gaseous material. Is it rather that another galaxy was created via Andromeda's supermassive black hole? Please explain."

"Yes; it is another galaxy that originated from Andromeda. The other galaxy is a different semi-phantom galaxy whose outer barriers of its black hole come into view by earthly inhabitants," replies the ATI,TPE. It continues, "The second galaxy was created by the residual matter sloughed off from the Andromeda galaxy such as the excess energies and matter particulates in their location via the black hole. It was a semi-phantom anomaly."

I added, "Is the second galaxy comprised of increased phantom matter but still semi-phantom for our ability to view it?"

"Per the All That Is, The Pure Essence: Yes; that is correct."

Supermassive black holes have ample speculation and study to imply an activity that can facilitate "new" or rather recycled creation, especially as wormholes. The All That Is, The Pure Essence describes the Milky Way supermassive black hole in response to my inquiries:

The supermassive black hole arose from a giant

supernova that was created extensively by the HU-4 Lyran Wars [originally in Galaxy-2's 12th dimension;[17] Bridge-A to the Milky Way was constructed to bypass critical destruction]. Its composition extends from the 11th dimension of the Milky Way down to the bottom tip of the 8th dimension in a linear direction twisting and turning along its abyss-like formation. It can extend for miles and miles but is interrupted with intrusion and penetration by outside forces and varied entity interferences. It can be formed along the way by additional, manipulated connections from outside forces and inside implosions.

The Milky Way supermassive black hole phenomenon began as a singular massive vacuum whereby a greater portion of the Milky Way galactic formation was built around its existence. It is currently a giant wormhole that was unnaturally created by agenda driven entities to enable mass dimensional groups to travel but not have the ability to stop at any particular point on this super highway.

The supermassive black hole super highways are connected by ramps to outside locations for entrance and landing in other worlds, galaxies, and universes. They act as thoroughfares, without ability to contain created platforms or allow for sustaining of life. Entities have endeavored to construct housing for existence in the black hole formations but have failed, even for temporary means and lengths of time.

I reveal that fractal mathematics and creations, including those built with the Fibonacci sequence, do not have infinite potential because they contain death science mechanics. I propose that eternal energy-matter and realms exist beyond our dense matter and outnumber the fragmented extensions.

Universal and cosmological creations are more diverse than physicists may assert with amalgamated unification theories of unified fields, large branes, or wave-particles containing the properties for all creation to come afterward. These "cosmic egg" ideas assume little or nothing of significance before them, which can equate to a religious God, although there are minor variations to the capacity that each paradigm attributes to its source. These perspectives are unfortunately tied to our experience of death science in which there is a beginning and an end, so they cannot extend beforehand to another paradigm of eternal science unless they open up to a very different perspective and physics. Our multifaceted abilities attest to experiencing a wide array of energies of which only some are commonly understood, but others could possibly never end.

We can diligently note patterns in the creation process without merging them into a super-creation or super-creator. I propose that each stage of cosmological creation merely provides an imprint for the next generation or level to develop somewhat differently, as the next section explains. Within a 15-dimensional galaxy, for example, a template originating in the 15th dimension would provide the information and ability for a patterned, step-down process that accumulates density in each subsequent large dimension with the potential for differences and distortions as it expands. Large dimensional creation is the result of an established process of outward expansion.

Origin and Expansion of Early Creation

It is not easy to gain precise information about the origin of creation because as entities, we provide our points of reference within our environmental and personal experiences. Information from a combination of entity groups can approximate the process of creation, but they cannot accurately explain the original essence and its creative process unless that original energy consciousness directly explains itself to them. I and my mother sense in our awareness that we have gained connection to the original, earliest energy consciousness. It can communicate directly to us and to all beings through our "inner knowing" when we diligently seek absolute purity. When I ask how it describes itself, it states, "All That Is, The Pure Essence."

As I explain in *Eternal Humans and the Finite Gods*, my mother asks the All That Is, The Pure Essence a question by which the end result is an answer in discernible, sparkling letters through her core, not her mind.[17] These letters are called fire letters, which are keylon light-symbol codes sent through our DNA in electromagnetic waves to produce sequentially arranged words.[20] My mother's uncommon ability operates largely in the subconscious state to translate the conscious questions to her "unconscious" inner body template. This pathway uses her delta wave frequency to link her to the first pre-field wave and then to the ATI,TPE. The ATI,TPE uses that same pathway to convey letters and words that are slowed down, so she can write them down. We also ask it follow-up questions to provide more detail. This process does not involve channeling or prophecy, which overrides the person.

Our pure intent can connect us to the All That Is, The Pure Essence, but our position in an expanded body that is partially comprised of fragmented structures does not make communication with the ATI,TPE entirely easy. Our unconscious or spiritual sense of "knowing" is most clear when we feel naturally and easily connected within our energetic self. Sometimes this knowing can reach further into our subconscious or even conscious state so we can mentally understand it. However, when putting it into the practice of technical precision, something may be lost in translation between the pure state of the ATI,TPE and our complex human composition. This is because the ATI,TPE is truly different from us. It only provides some degree of technical information, especially in non-eternal realms; our intuition and intelligence must do the rest.

My intuition has conveyed to me that pure pre-energy has one identity and is a single essence. Throughout my life, I knew that there was something ultimately simple and good somewhere, where complex entities and energies could not be. While sensing a pure conscious essence beyond us, my mother and I also feel a clear connection to purity in an innate part of ourselves that feels protected in its own energetic integrity. Upon discovering the All That Is, The Pure Essence and sensing a profoundly simple peace and good "energy" beyond our bodily, emotional, and mental influences, my mother and I continued to search but never find anything before it from both of our positions and what we can sense of its position. It just *is*, and it "feels" completely pure in my sensory awareness. Its title as All That Is, The Pure Essence is just that—a title or description that identifies itself as the simple but aware, pure essence that does not produce any vibration, frequency, sound tone, color, or distortion. It is not anything like the New Age misrepresentation as "All That Is," an omni-God Source containing every aspect of creation, which takes the original *All* describing the *Is* out of context and appropriates it to everything without discernment.

I asked the ATI,TPE via my mother's clear, internal connection for explanations to this question: "What is the All That Is, The Pure Essence?"

It explains itself and provides a simple flow chart:

> The All That Is, The Pure Essence is the origin of Itself, separate and boundless in its own space, nonmovable, and colorless. It is a point of pure conscious essence beyond all created matter, beyond the Void, beyond the separate still nothingness that comes after. It presents Itself as having primary knowledge with infinite conscious awareness.

> The All That Is, The Pure Essence first expanded Itself, not as desire but as Consciousness, when in a stationary state, by the intense, whelming pre-gaseous nature that comprises its pure essence. Its Consciousness intended to burst forth from that whelming state, laying pre-wave streams as it ventured forward and outward from its place of origin. This act was very simple and magnificent in its display,

similar to an instantaneous eruption.

The All That Is, The Pure Essence desires to explore and expand through the separate nothingness, through the endless Void to impart knowledge of its existence. This desire expands through spaces and between spaces via the bridge of Eia that creates as this desire flows. Anything beyond Eia, which carries the desire of purity and truth of the All That Is, The Pure Essence, has the possibility of distortion. The search for ultimate truth and purity enables the All That Is, The Pure Essence's desire as the extension for creation to increase the awareness of created beings.

All That Is, The Pure Essence
(*Stationary and conscious*) →
Separate film →
All That Is, The Pure Essence's extended
pure consciousness
(*Projected intention with the slightest
pre-wave movement*) →
Separate, still nothingness →
All That Is, The Pure Essence's extended pure desire
(*Desire with slight pre-wave movement expands
and explores measure*) →
Void →
Eia (*First pre-field wave*) →
Created matter (*Vibrational energy*) starting with
the "A" domain

The All That Is, The Pure Essence exists somewhere and somehow other than our experience. It is truly eternal with no end, and uniquely, no beginning. Something cannot arise from absolutely nothing, so it is conceivable that an exceptional "something" has always existed. As a conscious essence, it has intent that can gently expand as desire, flowing outward as a type of pre-wave not known by creation because it maintains its own space very close to its stationary origin.

The beauty of science is that it deduces what is "real" in existence, meaning that there is a type of substance in energy, even if the substance is ethereal and beyond our material experience. Consciousness and thought have a mechanism for action that extends beyond oneself, while the source of one's thoughts remains the same consciousness. We could erroneously imagine that the ATI,TPE and its intent and desire are all different substances because the original point as the All That Is, The Pure Essence does not move, while its intent and desire have slight movement; however, we know that the nature of consciousness is multifaceted. Thoughts build with intent and desire to eventually extend outside oneself, thus providing a foundation for communication and co-creation.

The unique, pre-gaseous "substance" of the ATI,TPE remains the only essence of the ATI,TPE. It refers to itself as a non-moving point, but it also states it is without boundary because its consciousness extends beyond that point with continuous creativity, as substantiated by its pre-gaseous nature. The ATI,TPE remains only itself in all situations because its energetic integrity is always preserved.

I asked the ATI,TPE, "Please explain what makes up the small layer of so-called 'film' that exists to separate the ATI,TPE from anything else?"

"Per the All That Is, The Pure Essence, the small 'film' is a translucent veil made up of the pre-atom essence's energy spark, spanning the endless nothingness."

The ATI,TPE is an eternal, pre-hydrogen-like pre-gas that endured the slightest hint of nuclear fission and fusion to expand itself. The ATI,TPE states that it desired to expand in order to connect with new creation outside of itself. Its intent caused slight internal pressure to begin that expansion; however, ATI,TPE does not approximate any force originating with the Big Bang. I suppose we can see everything in measures of familial substances, but this does not negate the significant differences between each stage of creation. Sometimes the original essence is entirely different from what is generationally created, such as with the ATI,TPE that does not share any substance with anything else in existence.

To clarify, the All That Is, The Pure Essence is the simplest essence in substance that is completely stationary as its foundation. Its nature contains the ability to minimally act in stages, not as a type of evolution in our estimation but as the inherent ability of consciousness to grow in awareness and expand beyond itself. Therefore, it became whelmed within itself, which caused an internal fission of sorts that did not divide itself; it extended its essence into slight movement as pre-wave streams. The intent and desire are secondary and tertiary aspects to its original stationary state, but they are all part of the same conscious essence as an identity and complete, pre-gaseous substance. The stages of its expansion include a slight film to provide a separation between its developed aspects. The ATI,TPE would still be complete unto itself if it never had expanded with intent and desire, but now with its outreaching extension, it has fulfilled its own potential that is built into its original essence.

Consciousness is a fascinating topic to explore in relation to existence and substance. When we apply these concepts to the ATI,TPE's original essence of consciousness, they are inherently united. While existence is the state of being, both consciousness and substance have the potential to grow. Consciousness has the most far-reaching ability, for it increases the awarenesses of self and the great beyond without limit.

People might describe the ATI,TPE as the origin and core essence of creation; however, it expresses that it is not a creator, nor does it actually reside within any creation. I asked it to further describe itself, to which it replies:

> The All That Is, The Pure Essence's desire to expand and intent to do so leads into the wording of a creator, but really it is not. It would be more accurate to state the following: It is the origin point, and through its intent and desire resides just outside

the core essence of creation.

The All That Is, The Pure Essence and its desired, created formations reside side by side. Its desire as transmitted by Eia connects to the core of each part of created matter, including core cells of the human body. The All That Is, The Pure Essence's desire seeks connections with pure intended created matter, and permeates that which desires the same in its consciousness.

The Void is a unique place because it was discovered by the process of expansion from the ATI,TPE to before the first pre-field wave. The ATI,TPE states that there is no energetic movement or consciousness in the Void; it is not a creative place. "The Void is subsequent endless space without the ability for creation. The All That Is, The Pure Essence did not desire creation there. The All That Is, The Pure Essence's desire was still expanding," it explains. The Void is not a field; it is a passageway.

The ATI,TPE expounds that the Void is a boundless space that initially held the potential for creation; however, it became an anomaly that does not contain any spark or creation. In a philosophical and technical perspective, the Void is nonexistent because it contains nothing; however, if there is an entirely empty space, then perhaps it somewhat exists. The original Void outside of creation is different from the spaces having no creational components and qualities within a creative domain, such as the universal spaces held in and between kathara grids.

Eia as the first pre-field wave is singular in consciousness that was created in and for its level of existence. It is very similar to the ATI,TPE's extended desire except that Eia's essence has a bit more wave-like ability and complexity. "Eia is the energetic bridge for the All That Is, The Pure Essence's existence which is instantaneous in connections to core aspects of creation," states the ATI,TPE.

Eia extends the desire of the ATI,TPE by creating the fastest and most "exceptional" vibratory movements and frequencies that exceed a trillion hertz, according to ATI,TPE. (The ATI,TPE explains that early waves are not powerful in the way of our density's four fundamental forces; it prefers to state that the "waves exude exceptional energy.") For a long time, I called this energy "Love" with a capital "L," but I finally decided to ask the ATI,TPE what its proper name or tone encryption is: Eia, pronounced I´ă (verified by Eia).

The ATI,TPE states that Eia is the first creator. The ATI,TPE does not consider itself a creator because it lacks the force of vibrational frequency. Regardless of this distinction, Eia is not an omnipotent creator (there is no such thing) because Eia is similar to the ATI,TPE in pre-gaseous essence, and it also has its own identity and space of separation between it and everything else. The reality of separation is built into every template of creation in order to preserve our individual identities and essences.

When my mother initially connected to Eia, she saw a faint cloud-like fog that expands. She elaborates:

> While I was communicating through my inner core with the All That Is, The Pure Essence and Eia, I saw an amazing bridge of energetic particulates appear as the frequency, sparkling within a fog-like gaseous cover from Eia to my true self. This is the expanded visual communication process within my human state.

This experience was my mother's first conscious connection to Eia, which she has since fine-tuned and no longer perceives any fog-like hue. At that time, I followed up with Eia to receive more understanding because I thought it should not have any color, including white, because it exists prior to light. It replied to my mother on my behalf:

> Per Eia, when "bridging" from the All That Is, The Pure Essence position in connection to creation, there is no color until the connection on the created

end is reached and percepted by the created recipient.
Colors are percepted where light penetrates matter.

Each act of creation provides an imprint or template imparted by the previous level, and that imprint becomes actualized as a new conscious substance. Eia's wave essence contains the imprint for our individual consciousness that I call one's "highest self," which corresponds to the next wave level in the earliest stage of the "A" domain. The highest self level is the furthest stage of ascension that we may embody as a waveform if we choose to transfigure that far. It is eternally and distinctly separate to the ATI,TPE, but it profoundly connects with the ATI,TPE unlike any other aspect of an entity. The highest self is one aspect of our entire conscious identity that is formed when we are first born or created as our respective higher or original self.

The term "higher self" refers to an entity's original conscious identity and energy-matter "body" from any higher dimension or cosmic level preceding the "lower self" human existence. Humans having a higher self have received a portion of the higher self's consciousness and imprint of its physical essence to form the respective human's foundational consciousness and energy-matter potential; this occurs a split second before the human body would integrate its own template's consciousness.

The higher self entity has left its location of origin to travel to Earth and reside in a nearby dimensional space to its lower self, confirms the ATI,TPE. The higher self's consciousness can be blended with its lower dimensional, human consciousness when the human is intentionally aligned with one's innate, true self. Conversely, the higher self's consciousness can be separated from the human's false sense of self and negative alignment with intervening entities.

New Age teachings misrepresent the term "higher self" to promote the one-but-numerous Law of One paradigm that diminishes individuality. Instead of having one's own higher self, or possibly no distinct higher self, it is shared among multiple, higher dimensional entities believed to be one's soul group.

To expand upon my definition for the original beings who are Earthlings, their DNA blueprint manifests similar but sometimes less advanced, higher dimensional aspects that do not come from another being. Technically, these humans do not have a higher self, but they do contain the same or similar imprints of energy-matter, which are not yet partially or fully actualized.

What religions call the "soul" could refer to the higher self or the higher self's energy-matter template within the human body (which is not the complete inner human essence), but the definition must be clear because they are distinct compositions. While our foundational, higher self component came from the higher self entity, it is ours within our Human structure. Thus, if we call our conscious, etheric human composition continuing onward after death a soul, then this "body" composition can transfigure with its DNA template to higher dimensions, matching their lighter densities until one's intrinsic mechanism can fully interface and blend the proper composition with one's higher self.

The higher or original self is a complete unit whose DNA transfiguration potential can reach the Cosminyahas level in the "A" domain. The highest self contains the ultimate potential of our conscious development beyond the DNA template. As the highest self is another aspect of one's consciousness, it should be partially activated or utilized wherever one is in creation.[17]

Another important aspect of our consciousness is our "core self" that provides the foundation for our DNA template and resultant embodiment. One's core self correlates to the second "A" domain level (after the highest self level), and due to its foundational nature, it energetically equates to the first DNA strand. An entity's core, which includes the "core self" and an imprint of the "highest self," is innately in all creations in subsequent levels, including simple creations as atoms and cells.

The first type of krystar arose as a primarily etheric, pre-gaseous composition with minimal pre-plasma. A krystar is an eternal type of star that can also provide planetary living conditions for a plethora of life forms. The first krystar is not sufficiently formed to house other entities, states the ATI,TPE; its waves exist in the earliest level of the "A" domain and energetically equate to one's highest self. Subsequent levels in the "A" domain contain increasingly developed krystar components.

The first inhabitable krystar exists in the Cosminyahas level, (pronounced Cŏsmeen´yahs), which is the last level of the "A" domain, reveals the ATI,TPE. The Cosminyahas krystars are composed of increased pre-plasma having excited, pre-ionized gas.

In Ashayana's process of seeking eternal creation from our very expanded position, before she heard about the Cosminyahas, she was informed about seven later krystars and Ah levels including Ah'-yah and Ah-yah-YA', which she calls the "Infinite Eternal Lands of 'Aah'" (p. 80).[62] She teaches that these areas are in an early plasmic realm called the CosMa'yah before the Eckasha-Aah; however, her Lands of "Aah" identification appears to place them in the Eckasha-Aah domain. The spellings of the CosMa'yah and Ah fields show that they are part of the Eckasha-Ah domain, which the ATI,TPE confirms, but it is not a rule to spell names according to their locations.

The seven sequential frequency tones Ka Ra Ya Sa Ta Ha La, introduced in the "Fractals and the Fibonacci Sequence" section, emanate from seven distinct levels containing krystars in the CosMa'yah realm. The Krysthal spiral stream originates in the first level of the Eckasha-Ah domain and CosMa'yah realm, states the ATI,TPE, and it contains the imprint of the pre-Krysthal spiral frequency band as its core originating in the Cosminyahas level.

The pre-Krysthal spiral frequency band continued down to the Eckasha-Ah domain where it formed a stream of energetic particulates, whose stream formation then gained increasing density appropriate to the foundation of latter domains. "In the beginning with the Krysthal spiral formation, it sparked its particulates in a fan-like fashion in multiple directions and then came together forming one precise path towards and through subsequent domains as an extension of its initial formation pattern," explains the ATI,TPE.

There are numerous Krysthal spirals streaming in specific creational pathways in realms within the domains. With my intuition and the assistance of the ATI,TPE, I understand that the Krysthal spiral joins other fundamental elements and aspects of eternal creation to develop new realms as creation branches out into subsequent domains. This process is similar to the image of a jellyfish with one or more umbrella-like tentacle extensions that assist the creation of one or more newly formed jellyfish, sometimes with an increasing number of "tentacle" branches.

When Ashayana learned about the Cosminyahas, she called it the "Sun-8" krystar, claiming it is the hidden but larger "Core" that surrounds and contains the seven later CosMa'yah levels represented as Suns 1-7, and also subsequent creation (p. S1 5-6).[71] This simply cannot happen because the seven later levels are in a separate domain than the Cosminyahas, and krystars, like stars and planets, exist in their own distinct spaces. However, the Law of One belief eliminates these boundaries in its concept of a giant gestalt sphere, which in this case is a Cosminyahas krystar. To the contrary, the Cosminyahas provides building blocks for the subsequent level's krystars. There is no giant sphere encapsulating everything as our source for life, because each krystar, entity, and etcetera is fundamentally self-contained.

As the original "point" of existence, the ATI,TPE does not conform to a spherical shape, nor does it encapsulate anything. It communicates its desire "permeates that which desires the same in its consciousness," which the following example can help illustrate.

There is a popular story about a philosophy teacher who demonstrates spatial capacity by using different media to fill an empty mason jar.[72] He first fills the jar with golf balls until no more can fit. Our body can be the jar, and the golf balls can be our molecules and atoms. Second, he inserts small pebbles that fill up smaller spaces. These can be subatomic particles. Third, he adds a bag of sand that can represent Krysthal scalar takeyons. Lastly, he pours in two beers. This is a large amount of liquid that can represent scales for silicate and plasmic energy-matter back to early creation. The jar contains one big space, but as it fills with different media having their own composition and space, the big space ends up containing spaces within or between spaces. The metaphor of a synaptic connection can apply when communicating energy and information through every separating film.

Outside of the jar is a formless essence as the All That Is, The Pure Essence existing in a separate space that can extend through the jar near the innermost aspect of each substance. Eia exists in a similar way to the ATI,TPE desire, facilitating a pure connection to energy-matter. Therefore, the ATI,TPE and Eia can connect with our pre-gaseous, pre-plasmic, "highest" (earliest) and core consciousness, which are aspects of the figurative beer that instantly flows to the rest of our naturally embodied attributes. It is through our foundational aspects that the ATI,TPE consciousness, extended by Eia, can reach our human existence when we are energetically aligned with pure intent. Our body as the jar inevitably feels the inside-to-outward flow.

The All That Is, The Pure Essence explains in more depth its ability to connect with creation:

> The All That Is, The Pure Essence as original point of existence extends its intent and desire by Eia enabling its energetic flow with Eia's conscious intent to create alongside actual creation. The All That Is, The Pure Essence with Eia do coexist with creation but do not mix with the creational connections provided by the natural levels and imprints within life forms.
>
> The All That Is, The Pure Essence remains pure, allowing minute reconfiguration in its composition since its original point of existence. This reconfiguration of composition enabled its conscious effort to extend its intent and desire to issue forth. Its energy and consciousness flow by Eia parallels to

creation. It connects to life forms including humans by aligning Itself with the life form's conscious, pure desire and intent for truth and knowledge within its inner constructs including DNA.

I reiterate that All That Is, The Pure Essence is not an entity or creation, and it does not consider itself a creator, although it does have creative intent and desire. When we attempt to define the origin of creation, sometimes we do so through our perceptual and experiential filters or rigid beliefs. For instance, we can attribute different definitions to the same word, such as *creator*:

1. Creator is a building block upon which similar but different energy-matter emerges.
2. Creator contains families of energy-matter that create similar extensions of those families.
3. Creator contains the same matter and forces imparted to all creation, and creation represents portions of that complete reservoir.

Numbers 1 and 2 are scientifically realistic while number 3 is a hypothetical belief. Number 3 involves the religious belief that states God is an immaterial and material everything from which we received our entire existence. This is the cosmic egg perspective that implies this ultimate entity is also an ultimate power as The Creator. Since there is a small separation between ATI,TPE and the core of every creation, and the ATI,TPE does not comprise any component of creation, it is impossible for the first stage of existence to contain all forces and material in a unified field.

The ATI,TPE and Eia are aligned due to their close similarity, much like how later creations share a similar resonance. This familial connection can employ a seamless collaboration, which many consider as a type of oneness, but the appropriate word is *togetherness*. I give caution to using the term *oneness* because it is primarily associated with the Law of One religion that desires to ultimately blend everyone into homogeneity.

The "togetherness" or coexistence of Eia and the ATI,TPE justifies the definition in number 1. After Eia's type of pre-field wave are creational layers with increasing amounts of energetic components and variations. When creational complexity involves similar but different identities coexisting agreeably in a realm or universe, this stage is number 2.

The Law of One belief is much more complex and contradictory to number 2; rather, it supports number 3. It states that we as very complex entities are merely different expressions of God Source wherein we embody all aspects of this Source but as different probabilities. This is pantheism at its finest, saying that everything is God. We are supposedly hologram replicas in different configurations of matter that an undivided field contains as "God," but then there are some aspects of division in that field in order to maintain a similar but diverse, "unified" field. The popular description of this God by New Age entity groups, especially the Infinite Awareness with author David Icke, is "All That Is, All That Has Been and All That Ever Can Be."[73] This completely misrepresents the actual All That Is, The Pure Essence, again showing the belief of entities instead of the ATI,TPE explaining itself.

Holograms are created when illuminating an object with a laser beam that has been split into two identical beams. Half of the beam is directed at the object, after which some of its reflected light is recorded onto a medium such as a photographic plate or photographic film with added light-reactive grains to increase resolution. The other half of the beam, the reference beam, is directed at a different angle with the use of a mirror toward the recording medium to recreate a virtual image from reconstructed and refracted wavefronts.[74] Frequency refraction, which bends and changes the direction of light and sound waves, is what helped create mostly similar but unequal copies of energy-matter in progressively altered galaxies away from Krysthal galaxies including AquaLaSha.

Physicist David Bohm observed that an image illuminated by

a laser and placed upon a photographic plate shows a holographic replica in every region of the plate. He proposed a cosmology stating there is an implicit order that should be traceable through every dimensional layer as "undivided wholeness in flowing movement" (p. 14).[75]

Michael Talbot, author of *The Holographic Universe*, provides a synopsis of Bohm's perspective:

> If a hologram of a rose is cut in half and then illuminated by a laser, each half will still be found to contain the entire image of the rose. Indeed, even if the halves are divided again, each snippet of film will always be found to contain a smaller but intact version of the original image. Unlike normal photographs, every part of a hologram contains all the information possessed by the whole. The "whole in every part" nature of a hologram provides us with an entirely new way of understanding organization and order. For most of its history, Western science has labored under the bias that the best way to understand a physical phenomenon, whether a frog or an atom, is to dissect it and study its respective parts....
>
> Bohm believes the reason subatomic particles are able to remain in contact with one another regardless of the distance separating them is not because they are sending some sort of mysterious signal back and forth, but because their separateness is an illusion. He argues that at some deeper level of reality such particles are not individual entities, but are actually extensions of the same fundamental something.[76]

Bohm's perspective supports the Infinite Awareness belief that states everything in our current experience is an illusion, and there should be no separation.[17] This belief can go so far to assert that our conscious, electromagnetic mind is an all-inclusive supercomputer creating each of our realities, and our imaginative interpretation of reality is a vibratory projection of this consciousness expressing itself through different reference points of spacetime. Therefore, there is no real individualized matter, just a multifaceted, superconscious energy. This viewpoint is what is illusory to me. It contradicts itself because it involves subjective reference points that are not identical holograms, thus implying individuality and separateness.

Bohm addresses this apparent contradiction when stating "relative independence" in his analogy of a flowing stream:

> On this stream, one may see an ever-changing pattern of vortices, ripples, waves, splashes, etc., which evidently have no independent existence as such. Rather, they are abstracted from the flowing movement, arising and vanishing in the total process of the flow. Such transitory subsistence as may be possessed by these abstracted forms implies only a relative independence or autonomy of behaviour, rather than absolutely independent existence as ultimate substances (p. 62).[75]

Bohm and I agree to an extent about a higher and inner dimensional status as an inherent order that projects an extension of itself in our currently "objectified" existence. However, we differ where his view takes on a spiritualism that treats everyone's subjective experiences as variable experiences of one unit. This somehow implies we are selfsame extensions of the original consciousness, thereby affecting it equally as it affects us.

While many extreme oneness proponents believe the ultimate God state is the non-moving zero-point, Bohm considers this state and the condition of separateness to be static and fragmented. He states, "[M]y main concern has been with understanding the nature of reality in general and of consciousness in particular as a coherent whole, which is never static or complete, but which is in an unending process of movement and unfoldment" (p. x).[75] He bases his belief on the annihilation model in which energy-matter continually changes, which is why he uses the analogy of the physical stream with its interactive components for our proposed unified field of existence. His view turns us into a moving nebula or storm of instability that projects an image of separateness and stability in our reality. Accordingly, he claims that the mind and body are projections from this higher dimensional, undivided existence, which implies a single entity. In his computations, Bohm assumes "non-denumerable infinity of variables" from fractions, not whole numbers, thus proving his fractal-based inclination (p. 117).[75]

We typically define something non-moving as static, but this is not necessarily an equal association. *Static* has a negative connotation of being unchanging as if stubborn, which shows an incomplete or fragmented nature. It also applies to electricity that is imbalanced in its charges and stuck in place. Again, we are limiting our logic and understanding to our respective energy-matter when imagining something that simply exists in a stationary way. The All That Is, The Pure Essence is the essence of life that is complete unto itself, not fragmented in any way.

The saving point of Bohm's argument is that he opens up scientific thought toward pre-Big Bang conditions and quantum potential in a super-implicate, structural order. However, his theory about an infinite series of representations implied by ceaseless enfoldment and unfoldment, which would constantly dissolve subatomic particles into the implicate order and then recrystallize them, necessitates death by fractal geometry limiting and deforming the infinity. The ultimate Fibonacci or similarly based fractal is doomed to space dust or a compressed black hole, so it cannot reach the super-implicate level. I see his belief in the Law of One in his astute attempts to connect modern physics to metaphysics. I invite progressive, niche physicists to further climb outside the religious box and explore the information that I and Krysthal Science provide.

Creations are essentially familial microcosms of their respective macrocosms because we cannot make up something out of nothing, but none is a hologram or replica of All That Is, The Pure Essence, Eia, or any Krysthal stage of development. Individual identity is primary, and expansion occurred through the desire to connect and love; this is the foundation of our makeup. We are built to be eternally individual and loving. We never, ever lose these foundational aspects of ourselves unless we no longer would have the desire to exist, and more importantly, if we would detest existence. Even very evil entities still wish to exist, but they are hanging onto life by a thread because they detest natural life in some regard. If they sever their connection to the ATI,TPE in the possible event of shattering to space dust, nothing happens to the ATI,TPE in the process. The ATI,TPE is never divided or outside of itself, so it merely stays in its full integrity.

Ashayana has spoken about the possibility arising in which the identity imprint of the dissolved person is preserved in an early plasmic level (of the Eckasha-Ah) called Ah-YA'-yah, where the imprint can merge with other creational aspects to be literally born again.[62] This self-identity construct would have its past distortions removed, although I doubt that all will have been forgotten because the collaborative nature of one's being contains a measure of consciousness and awareness in every level of itself. This reconstructed identity does not equate to the pre-plasmic wave essences of one's highest and core self that form the foundation of entity consciousness.

Two levels before Ah-YA'-yah is the Ah'-yah level that Ashayana described as the essence of the "massive consciousness field, that living consciousness field of Source" (DVD-1) and "'The Essence of' the Unutterable ONE" (DVD-2).[62] Since stating those levels of creation, she has learned about the Cosminyahas from her entity-based transmissions, which may give room in her teachings to accept that individualized identities can exist as distinct pre-gaseous, pre-plasmic, eternal substances in a level preceding the earliest inhabitable krystar; however, this would require a change in belief to accept that individuality is real. Eia is also not yet discovered or revealed by her entity groups, but it is relatively not far from the more complex Cosminyahas level.

Unfortunately, religious-minded entity groups often lead us to believe that each new revelatory level is the God Source they have been seeking when it is still part of creation, and they may later realize—if they probe with pure intent—that there is something earlier. I suggest they change their perception about any God and hierarchy to stop misrepresenting unique levels of existence.

The partiki light-body template cannot exist without the inner pre-gaseous and pre-plasmic templates. Our earthly body is an excellent example because despite its distortions, we can still communicate in some way with our inner template all the way back to the All That Is, The Pure Essence by way of Eia, which can act as a bridge to override our disconnected parts. I "see" in my awareness that our body's inner units, waves, and fields are like preserved, laminated layers in perhaps a type of meshed structure within each light-body layer, and several fundamental components are between the light-body layers. This allows each familial structure to coexist, where the "A" waves connect with the Ah versions of pre-gas and pre-plasma, which connect with the latter light-body versions of partiki generations. Every actualization of these imprints entails a unique variation of the measurably complex family. The ATI,TPE states that partiki clusters, for example, do not necessarily replicate the same number of partiki in each unit's process of expansion. We are not tied to a field or structure that came before us, but we do carry aspects of their natures within us.

The unified field concept appears to be muddled, so I will bring simple reasoning and information into it. All natural levels of creation, including their respective entities, are complete in themselves. None of them contains every component or property of creation, but they do contain their own "source" or origin of some familial entities that thrive in their own environment.

The 20,736 partiki Yunasum unit, for example, can communicate with other partiki units, but it will more easily communicate with other Yunasum, collectively called Yunasai.

Each Yunasum is in a separate but most similar location across creation compared to other partiki units, potentially providing a type of field or group consciousness for the Yunasai while not assimilating their individual members into a single shape, blob, or gestalt. I cannot emphasize enough how creation contains multiple fundamental aspects, so a single law cannot adequately explain them unless it truly involves everything as they individually are in respective locations.

In the MCEO-GA's Law of One belief where cosmic structure contains spheres within spheres of cycles, each subsequent domain and layer of creation spins one direction (such as the Eckasha) and creates through deflection a counter-rotating set of spheres (such as the Ecka).[77] This model severely simplifies and misrepresents creation by stating that the cosmos replicates itself via inherent forces into smaller and smaller fields, stars, planets, and species within its original star-like boundary.

The nature of life is truly limitless, as the ATI,TPE concurs and conveys, "Natural creation is limitless because the All That Is, The Pure Essence is limitless." Many scientists do not yet understand that fractal and Fibonacci-based creations cannot provide unlimited expansion, but they do witness eternal creation mechanics existing in our universe that provide life, regenerative health, and balance between separateness and togetherness.

Early creation after Eia contains eternal codes for all subsequent Krysthal creations, but these codes are specific blueprints until the levels actualize them in their respective ways. Humans contain these blueprints as well as other codes, so we are sufficiently whole although our complete abilities are not yet actualized. Accordingly, Galaxy-2's 15th dimension contains all the fundamental codes of the Milky Way, and all Milky Way dimensions, including the 1st dimension, contain these same fundamental codes. The differences between creational codes in earlier and later domains, realms, and levels are their amount and diversity, especially when something is added from a phantom state, because of the expansion process typically accumulating more components. Fully eternal entities cannot contain phantom codes because phantom energy-matter comes afterward, usually in different realms; however, these entities can contain the potential to become compromised. Typically, a distorted reconfiguration occurs after one's energetic discernment has become lost, which stresses the importance of keeping clear demarcations in both science and reason.

Cosmology is an enormous subject that becomes limited when we seek to define all energetic substances by a scientific law of imagined simplicity. I suppose we could loosely define a large dimension as one or more unified fields, but there are other components built into every large dimension to suggest that we no longer equate a unified field or law to an entire universe. The concept of a unified field or law loses even more credibility when equated to a vast cosmos.

The cosmology I present begins with the All That Is, The Pure Essence and includes the unique existences of every conscious substance that naturally interact in harmony, as well as the lesser amount of unnatural diversions in fragmented creations. Energetic harmony builds bridges and emanates a type of in and out "breath" flow facilitating the processes of creation and potential ascension and descent to familial realms and energy-matter. Similar to the natural dimensional ascension process toward the Ecka, the transfiguration from outer to earlier domains can be easy. In fact, as our dense matter becomes more pliable and plasmic, boundaries are less detectable, making it easier to be alongside Eia and the ATI,TPE.

Conclusion

Consciousness exists in all living things, and things of all levels incorporate a type of substance. Conscious awareness starts the process of creation; therefore, all living things can essentially create. We can perceive the first layer of substance as the most intelligent because it started the process upon which we depend and connect to subsequent layers, but playing the game of "Who or what is better?" can erroneously negate the innate intelligence of every other substance knowing how to live in its environment.

Science and religion alike tend to emphasize an origin or creator as more capable than what ensues in creation, where it seemingly owns the patent of ultimate design, and what follows can only imperfectly approximate its omnificence. Although mainstream spirit-science is dismissed as pseudoscience by most scientists, it has infiltrated mainstream cosmology in the attempt to merge complex realities into an oversimplified anomaly as a multifaceted particle, field, or brane providing all energies and matter. I argue that a truly simple source can exist when we open up to a much larger cosmology between our respective locations with different layers, scales, substances, interactions, and spaces between spaces. Physicists may intuit a simple origin as the All That Is, The Pure Essence, but they are hindered by their scientific method that ties us to the parameters of our highly expanded, physical reality.

The All That Is, The Pure Essence does not approximate any of our substances, even if we may loosely describe it as point-like or pre-gaseous. There are far too many compositional stages between the ATI,TPE and hydrogen to assert any correlation. Its pre-Krysthal "pre-hydrogen" essence is an exponential "pre-" so loosely related to hydrogen that they do not comprise any similar substance. Any wave, field, or plasma as we know it also does not correlate to Eia.

Cosmological aspects can have a connection and slight relation to us in the big picture, but we should endeavor to state things as they are, where they are, no matter their relation to us. This is not an easy task, which is why our awareness must expand beyond ourselves, while also not losing ourselves.

If we might attribute a perfect design to the ATI,TPE, it would be ultimate, eternal purity and simplicity with inherent uniqueness that reaches out to facilitate the existence of more eternal, unique purity. It is beyond any power, control, or god-like ego that has become embedded into our minds by untoward, otherworldly entities touting themselves as Gods who have lost sense of their core's purity. Where there is power and hierarchy, there is usually subjugation; together, they perpetuate the victim-victimizer game that has stifled humanity's greater awareness and self-confidence to push past destructive energies and mentalities.

Our current position among a seemingly dichotomous mix of life and death gives us the opportunity to scrutinize their differences, thus organizing them into their respective creational positions instead of muddling them into one reality. Instead of evaluating science in exclusion to religion or vice versa, I propose that we put all sciences and religions on the table to determine how personal bias has potentially entered into scientific hypotheses and theories of cosmology, and how we can arrive at a more accurate picture.

Although the ATI, TPE and early creation are vastly different from our complex reality, we do not need to employ miraculous thinking to describe the origin of existence as religion tends to do, and where physics has joined it. Instead, we can view each creational process as a building block or springboard toward a new creation. The cosmology I propose is enormous with multiple pathways and lifelines of energetic integrity that help sustain us as beings when relatively new death science mechanics affect our universe and body.

Science and religion can agree about intelligent creation. Partial or fully organized patterns exist behind every independent or random occurrence, and the pre-matter substances of those patterns are familial entities with similar but distinct consciousness that communicate among themselves; in addition, they communicate with distant generations when there is co-resonance. Unfortunately, science and religion are most aligned in their acceptance of fractal mechanics because the Ecka-Veca-based entities who originally created religions are the same ones who actively fabricated death science, including the supermassive black hole matrix. I think it is now time that we ground our imagination in reality by allowing an even playing field for all of our senses and abilities instead of dismissing some in favor of a perceived hierarchy of importance; these gentler energies within and beyond us are just as real as the denser ones that dominate our attention.

Bibliography

1. **Klyce, Brig.** Comets: The Delivery System. *COSMIC ANCESTRY.* [Online] [Cited: April 2, 2016.] http://www.panspermia.org/comets.htm.
2. **NASA/WMAP Science Team.** How Old is the Universe? *NASA: Universe 101.* [Online] National Aeronautics and Space Administration, December 21, 2012. http://map.gsfc.nasa.gov/universe/uni_age.html.
3. **Vanderbilt University.** Inflationary Cosmology on Trial. *YouTube.* [Online] YouTube, LLC, April 4, 2011. Video; 1:23:56. http://www.youtube.com/watch?v=IcxptIJS7kQ.
4. **Howell, Elizabeth.** What Is the Big Bang Theory? *SPACE.com.* [Online] Purch, June 12, 2017. http://www.space.com/25126-big-bang-theory.html.
5. **O'Callaghan, Jonathan.** Dark Matter is Being Measured More Accurately Than Ever Before. *IFLSCIENCE!* [Online] July 10, 2015. http://www.iflscience.com/space/embargo-9-july-1500-bst-dark-matter-being-measured-more-accurately-ever.
6. **Redd, Nola Taylor.** What is Dark Energy? *SPACE.com.* [Online] Purch, May 1, 2013. http://www.space.com/20929-dark-energy.html.
7. **Musser, George.** According to the big bang theory, all the matter in the universe erupted from a singularity. Why didn't all this matter--cheek by jowl as it was--immediately collapse into a black hole? *Scientific American®.* [Online] Nature America, Inc., September 22, 2003. http://www.scientificamerican.com/article/according-to-the-big-bang/.
8. **Villanueva, John Carl.** Big Freeze. *Universe Today: space and astronomy news.* [Online] Universe Today, April 26, 2016. http://www.universetoday.com/36917/big-freeze/.
9. **NASA/WMAP Science Team.** What is the Ultimate Fate of the Universe? *NASA: Universe 101.* [Online] National Aeronautics and Space Administration, June 29, 2015. http://map.gsfc.nasa.gov/universe/uni_fate.html.
10. How is a Black Hole Created? *HubbleSite.* [Online] Space Telescope Science Institute. [Cited: August 19, 2017.] Information distributed under contract with NASA. http://hubblesite.org/reference_desk/faq/answer.php.id=56&cat=exotic.
11. How do scientists measure or calculate the weight of a planet? *Scientific American®.* [Online] Nature America, Inc., December 12, 2005. Originally published on March 16, 1998. http://www.scientificamerican.com/article.cfm?id=how-do-scientists-measure.
12. **Cornell University Department of Astronomy.** Black Holes and Quasars. *Ask an Astronomer.* [Online] The Curious Team. [Cited: February 28, 2014.] http://curious.astro.cornell.edu/the-universe/black-holes-and-quasars.
13. **Plait, Philip.** Black Holes: From Here to Infinity. [Online] [Cited: June 30, 2017.] http://www.spitzinc.com/pdfs/educ_guide_blackholes_nasa.pdf.
14. Quasar. *Wikipedia, The Free Encyclopedia.* [Online] Wikimedia Foundation, Inc. [Cited: March 2, 2014.] http://en.wikipedia.org/wiki/Quasar.
15. The Universe. *Sloan Digital Sky Survey / SkyServer.* [Online] Astrophysical Research Consortium. [Cited: March 2, 2014.] http://cas.sdss.org/dr3/en/proj/basic/universe_original/.

16. **MCEO Freedom Teachings®**. *Spirals of Creation.* Azurite Press MCEO, Inc., 2010. Class module compiled by Kathara Team.

17. **Talea, Theresa.** *Eternal Humans and the Finite Gods: How an Ex-Prophet and I Left Religion and Discovered Universes Beyond and Within.* 3rd Ed. Rancho Cordova, CA : Rediscovery Press, 2017.

18. **MCEO Freedom Teachings®.** *The Kathara™ Bio-Spiritual Healing System; Level-1 Certificate Program.* [Handbook] Allentown, PA : A'sha-yana Deane, Azurite Press MCEO, Inc., 2000.

19. **MCEO Freedom Teachings®.** *Engaging the Load-Out; The Last Ascension Cycle and the Gate of AshaLA.* [DVDs] Phoenix, AZ : Azurite Press MCEO, Inc., Adashi MCEO LLC, January 2008.

20. **Deane, Ashayana.** *Voyagers: The Sleeping Abductees, Volume I of the Emerald Covenant CDT-Plate Translations.* 2nd Ed. Columbus, NC : Wild Flower Press, 2002.

21. **MCEO Freedom Teachings®.** *Kathara Levels 2 & 3 Foundations: "Awakening the Living Lotus": Healing Facilitation through Crystal Body Alignment.* [Handbook] Phoenix, AZ : Azurite Press MCEO, Inc., Adashi MCEO LLC, April 2004.

22. **Weisstein, Eric.** Tube. *Wolfram MathWorld.* [Online] Wolfram Research, Inc. [Cited: May 24, 2016.] http://mathworld.wolfram.com/Tube.html.

23. **MCEO Freedom Teachings®.** *Sliders-9: "Advanced Spiritual Body Training - The Flame of CosMAyah, Mayan Mother Matrix & Luminary Body Activation".* [Handbook] Azurite Press MCEO, Inc., Adashi MCEO LLC, October 2010 & January 2011.

24. **Bowman, Carol.** *Children's Past Lives: How Past Life Memories Affect Your Child.* New York : Bantam Books, 1997.

25. **Einstein, Albert.** Doc. 30: The Foundation of the General Theory of Relativity. [trans.] Alfred Engel. *The Collected Papers of Albert Einstein, Volume 6: The Berlin Years: Writings, 1914-1917 (English translation supplement).* Princeton, NJ : Princeton University Press, 1997.

26. **Eckardt, Horst and Laurence G. Felker.** Einstein, Cartan and Evans--Start of a New Age in Physics? *Alpha Institute for Advanced Studies (AIAS).* [Online] December 9, 2005. Translation of original article published in NET-Journal. http://www.aias.us/documents/eceArticle/ECE-Article_EN.pdf.

27. *Fair sampling perspective on an apparent violation of duality.* **Bolduc, Eliot, Jonathon Leach, Filippo M. Miatto, Gerd Leuchs, and Robert W. Boyd.** Iss. 34, Washington, DC : National Academy of Sciences, August 26, 2014, PNAS, Vol. 111, p. 12337.

28. *Spatially structured photons that travel in free space slower than the speed of light.* **Giovannini, Daniel, Jacquiline Romero, Václav Poto ek, Gergely Ferenczi, Fiona Speirits, Stephen M. Barnett, Daniele Faccio, and Miles J. Padgett.** Iss. 6224, Washington, DC : American Association for the Advancement of Science, 2015, Science, Vol. 347, pp. 857-860.

29. **McMahon, David.** *Quantum Field Theory Demystified: A Self-Teaching Guide.* San Francisco : The McGraw-Hill Companies, Inc., 2008.

30. **Fowler, Michael.** Maxwell's Equations and Electromagnetic Waves. *Galileo and Einstein.* [Online] Physics Department, University of Virginia. [Cited: December 15, 2015.] http://galileoandeinstein.physics.virginia.edu/more_stuff/Maxwell_Eq.html.

31. **Fermilab.** Particles, Fields and the Future of Physics - A Lecture by Sean Carroll. *YouTube.* [Online] YouTube, LLC, July 11, 2013. Video, 1:37:54. http://www. youtube.com/watch?v=gEKSpZPByD0.

32. **Pagels, Heinz R.** *Perfect Symmetry: The Search for the Beginning of Time.* Trade Paperback Ed. New York : Simon & Schuster Paperbacks, 2009. p. xvii.

33. Manifold. *Wikipedia, The Free Encyclopedia.* [Online] Wikimedia Foundation, Inc. [Cited: April 9, 2016.] http://en.wikipedia.org/wiki/Manifold.

34. **Maldacena, Juan.** Who's Counting? Is it 10 or 11? (dimensions, that is ---M Theory is making me Manic!). *IAS School of Natural Sciences.* [Online] [Cited: February 3, 2016.] http://www.sns.ias.edu/ckfinder/userfiles/files/Dimensions.pdf.

35. **McMahon, David.** *String Theory Demystified: A Self-Teaching Guide.* San Francisco : The McGraw-Hill Companies, Inc., 2009.

36. **Schwarz, Patricia.** Is there a more fundamental theory? *The Official String Theory Web Site.* [Online] [Cited: May 14, 2016.] http://www.superstringtheory.com/ basics/basic7.html.

37. **NOVA.** The Elegant Universe: Part 3. *PBS: Public Broadcasting Service.* [Online] WGBH Educational Foundation, November 3, 2003. Video, 53:07. http://www. pbs.org/video/1512308858/.

38. **Stein, Leo C.** How can a universe with 11 dimensions exist? *Quora.* [Online] Quora, Inc., October 17, 2010. http://www.quora.com/How-can-a-universe-with-11-dimensions-exist.

39. **Groleau, Rick.** Resonance in Strings. *NOVA: Science Programming on Air and Online.* [Online] WGBH Educational Foundation. [Cited: December 6, 2017.] http://www.pbs.org/wgbh/nova/elegant/resonance.html.

40. Planck Length. *Cosmos - The SAO Encyclopedia of Astronomy.* [Online] Swinburne University of Technology. [Cited: August 28, 2015.] http://astronomy.swin.edu.au/ cms/astro/cosmos/p/Planck+Length.

41. **Brown, Eryn.** First three-year LHC running period reaches a conclusion. *Media and Press Relations.* [Online] CERN, February 14, 2013. http://press.cern/press-releases/2013/02/first-three-year-lhc-running-period-reaches-conclusion.

42. **Dorminey, Bruce.** CERN's Higgs Discovery As Portal To New 'Technicolor' Physics. *Forbes.* [Online] Forbes Media LLC, November 19, 2014. http://www. forbes.com/sites/brucedorminey/2014/11/19/cerns-higgs-discovery-as-portal-to-new-technicolor-physics/.

43. **Thomas, Kelly Devine.** Discovering the Higgs: Inevitability, Rigidity, Fragility, Beauty. *IAS.* [Online] Institute for Advanced Study, Spring 2013. http://www.ias. edu/ideas/2013/higgs-arkani-hamed-maldacena.

44. **Inglis-Arkell, Esther.** There is no such thing as emptiness. There is only quantum foam. *io9: We Come From the Future.* [Online] Gawker Media, March 18, 2013. http://io9.gizmodo.com/there-is-no-such-thing-as-emptiness-there-is-only-quan-453814024.

45. Claud Lovelace. *Wikipedia, The Free Encyclopedia.* [Online] Wikimedia Foundation, Inc. [Cited: June 12, 2016.] http://en.wikipedia.org/wiki/Claud_Lovelace.

46. **Peat, F. David.** *Superstrings and the Search for the Theory of Everything.* Chicago : Contemporary Books, 1998. p. 58.

47. **Alexjander, Susan.** Microcosmic Music - A New Level of Intensity. *Our Sound Universe--the Music of Susan Alexjander.* [Online] [Cited: January 16, 2011.] http:// web.archive.org/web/20110930062656/http://www.oursounduniverse.com/articles/ microcosmic.html.

48. **Simonetti, John.** Frequently Asked Questions About Quasars. *Virginia Tech Physics.* [Online] [Cited: January 4, 2014.] http://www.phys.vt.edu/~jhs/faq/quasars. html#q11.

49. **Johnston, Hamish.** Antihydrogen trapped at CERN. *physicsworld.com.* [Online] Institute of Physics, November 17, 2010. http://physicsworld.com/cws/article/ news/2010/nov/17/antihydrogen-trapped-at-cern.

50. **Chung, Emily.** Antimatter atom 'measured' for first time. *CBCnews: Technology & Science.* [Online] CBC News, March 7, 2012. http://www.cbc.ca/news/technology/ story/2012/03/07/science-antimatter-alpha-hayden.html.

51. **Soffer, Abner.** What's the matter with antimatter? *Particle Physics in Plain English!* [Online] [Cited: May 10, 2015.] http://conferences.fnal.gov/lp2003/forthepublic/ matter/index.html.

52. **SPACE.com Staff.** Why We Exist: Matter Wins Battle Over Antimatter. *SPACE. com.* [Online] Purch, May 18, 2010. http://www.space.com/8441-exist-matter-wins-battle-antimatter.html.

53. **STFC Science in Society Team.** Does antimatter exist? *STFC Large Hadron Collider.* [Online] Science and Technology Facilities Council. [Cited: August 8, 2013.] http://web.archive.org/web/20130808070547/http://www.lhc.ac.uk/ The+Particle+Detectives/Take+5/13685.aspx.

54. Matter-Antimatter Annihilation Visualization. *echochamber xkcd: Forums for the webcomic xkcd.com.* [Online] June 6, 2010. Science forum post by PM 2Ring. http://forums.xkcd.com/viewtopic.php?f=18&t=61064.

55. Scalar Potentials and Scalar Waves. *Research Media & Cybernetics.* [Online] RMCybernetics. [Cited: January 25, 2014.] http://www.rmcybernetics.com/ science/physics/electromagnetism2_scalar_waves.htm.

56. **MCEO Freedom Teachings®.** *Festival of Light Celebration: The Starfire Cycle and the Re-birth of the Original Amenti Rescue Mission.* [DVDs] London, UK : Azurite Press MCEO, Inc., Adashi MCEO LLC, February 2006.

57. **Deane, Ashayana.** *Voyagers: The Secrets of Amenti, Volume II of the Emerald Covenant CDT Plate Translations.* 2nd Ed. Columbus, NC : Wild Flower Press, 2002.

58. **MCEO Freedom Teachings®.** *The Tangible Structure of the Soul: Accelerated Bio-Spiritual Evolution Program.* [Handbook] A'sha-yana Deane, Azurite Press MCEO, Inc., 2000.

59. **Bellis, Mary.** An Atomic Description of Silicon: The Silicon Molecule. *ThoughtCo.* [Online] About, Inc., March 6, 2017. http://www.thoughtco.com/atomic-description-of-silicon-4097223.

60. Tara, Goddess of Peace and Protection. *Goddess Gift.* [Online] The Goddess Path. [Cited: February 2, 2014.] http://www.goddessgift.com/goddess-myths/goddess_ tara_white.htm.

61. **Atsma, Aaron J.** Gaia. *Theoi Greek Mythology.* [Online] Theoi Project. [Cited: February 2, 2014.] http://www.theoi.com/Protogenos/Gaia.html.

62. **MCEO Freedom Teachings®.** *Sliders-8: Preparing the Body for Slide--Advanced Level. Awake, Aware, and ALIVE in the Lands of Aah: The "Sea of Ah'yah," Eternal Stream of Ah-yah-YA', the Covenant of Ah-Yah-RhU', and Eternal Dream-Fields of the ONE.* [DVDs and Handbook] Phoenix, AZ : Azurite Press MCEO, Inc., Adashi MCEO LLC, August 2010.

63. **The Aetherius Society.** Ascended Masters and the Spiritual Hierarchy of Earth. *The Aetherius Society: Co-operating with the Gods from Space.* [Online] The Aetherius Society. [Cited: June 30, 2017.] http://www.aetherius.org/ascended-masters/.

64. **MCEO Freedom Teachings®.** *The 12-Tribes Transcripts, Class 8.* [Handbook] Phoenix, AZ : Azurite Press MCEO, Inc., Adashi MCEO LLC, October 2007.

65. **Bernard, Raymond W.** *The Hollow Earth.* 2nd Ed. Pomeroy, WA : Health Research Books, 1996. p. 2.

66. **MCEO Freedom Teachings®.** *Sacred Sexuality & the Art of Divine Relationship: Sacred Sex, Divine Love & Eternal Co-Creation, Part One.* [DVDs] Denver, CO : Azurite Press MCEO, Inc., Adashi MCEO LLC, July 2006.

67. **Deane, Ashayana.** *Introductory-Topic Summary-2.* Azurite Press MCEO, Inc., 2009. Part of the MCEO Freedom Teachings® Series. Cited January 12, 2018, this report is reposted at http://lightworkers.org/channeling/91505/crucial-information-regarding-2012-guardian-alliance-through-asha.

68. **Melchizedek, Drunvalo.** Holy Mer:.Ka:.Ba:. Meditation. *The Blue Brethren.* [Online] [Cited: June 30, 2017.] http://www.bibliotecapleyades.net/bb/drunvalo.htm.

69. Properties of the number 144. *RidingTheBeast.com.* [Online] December 19, 1998. http://www.ridingthebeast.com/numbers/nu144.php.

70. **MCEO Freedom Teachings®.** *The Melchizedek Cloister Emerald Order (MCEO) Ordinate System: Getting Your Ascension Codes Back.* [Handbook] Azurite Press MCEO, Inc., Adashi MCEO LLC, 2007.

71. **The Tan-Tri-Ahura Teachings--The Path of Bio-Spiritual Artistry.** *Treasures of the Tan-Tri-Ahura: Gate-Walkers, Wave-Runners and Star-Riders of the Krystal River Host.* [DVDs and Handbook] Sarasota, FL : E'Asha Ashayana, ARhAyas Productions, AMCC-MCEO, LLC, August 17-20, 2012.

72. A Teacher's Life Lessons Using a Jar and Some Golf Balls. *sunny skyz.* [Online] CK Media Group, October 23, 2012. http://www.sunnyskyz.com/feel-good-story/111/A-teacher-s-life-lessons-using-a-jar-and-some-golf-balls.

73. **Icke, David.** Book Description (back cover). *Remember Who You Are: Remember 'Where' You Are and Where You 'Come' From.* Ryde, UK : David Icke Books Ltd, 2012.

74. **Workman, Robert.** What is a Hologram? *LiveScience.* [Online] Purch, May 23, 2013. http://www.livescience.com/34652-hologram.html.

75. **Bohm, David.** *Wholeness and the Implicate Order.* New York : Routledge Classics, 1980.

76. **Talbot, Michael.** The Holographic Universe: Does Objective Reality Exist? *rense.com.* [Online] March 12, 2006. http://www.rense.com/general69/holoff.htm.

77. **MCEO Freedom Teachings®.** *The Cosmic Clock Re-set: Entering the Reucha-TA Great Healing Cycle.* [DVDs] Phoenix, AZ : Azurite Press MCEO, Inc., Adashi MCEO LLC, September 26-28, 2003.

www.ingramcontent.com/pod-product-compliance
Lightning Source LLC
Chambersburg PA
CBHW020210200326
41521CB00005BA/325